Essentials of Navigation

Hongwei Bian · An Li · Heng Ma · Rongying Wang

Essentials of Navigation

A Guide for Marine Navigation

Hongwei Bian
Naval University of Engineering
Wuhan, Hubei, China

An Li
Naval University of Engineering
Wuhan, Hubei, China

Heng Ma
Naval University of Engineering
Wuhan, Hubei, China

Rongying Wang
Naval University of Engineering
Wuhan, Hubei, China

ISBN 978-981-99-5635-7   ISBN 978-981-99-5636-4 (eBook)
https://doi.org/10.1007/978-981-99-5636-4

Jointly published with Science Press
The print edition is not for sale in China (Mainland). Customers from China (Mainland) please order the print book from: Science Press.
ISBN of the Co-Publisher's edition: 978-7-03-074703-7

This Springer imprint is published by the registered company Springer Nature Singapore Pte Ltd.
The registered company address is: 152 Beach Road, #21-01/04 Gateway East, Singapore 189721, Singapore

Paper in this product is recyclable.

Preface

Navigation is a fundamental ability of creature. For humans, navigation technology determines their ability to accurately perceive motion, make path decisions, and control their own movements and those of their carriers. Being a fundamental requirement for human survival and social development, navigation, as one of the oldest scientific techniques in humanity, not only has a long history but also maintains constant and dynamic development today. The rapid development and tremendous impact of different navigation technology can be seen everywhere, from the widespread availability of mobile terminal positioning services to government-sponsored satellite navigation systems, from intelligent driving for civil use to high-precision positioning of space spacecraft and underwater vehicles in complex environments.

Navigation capabilities have become an overall indicator of national strength, greatly affecting our social life and military transformation in many ways. In the military field, the impact of precision guidance and navigation positioning technology has long risen from the tactical level up to the overall level in modern warfare. For example, positioning, navigation, and timing (PNT) capabilities have become one of the most important basic technical guarantees in modern warfare. Meanwhile, location-based services (LBSs) have also become one of the core foundations of new industrial indoor and outdoor Internet of things (IoT), active in the civil sectors such as logistics transportation, warehouse management, fire rescue, and shared mobility.

Navigation technology is gaining growing attention because of its wide range of applications. In terms of the basic functions, navigation needs to solve problems in three aspects so as to guide carriers to their destinations: to determine the motion parameters of the carrier, to determine the given environmental information, and to make route decisions and guide the carrier to the destination. The three types of technologies concerned are relatively independent and at the same time closely related, involving multiple fields such as surveying and mapping, detection, instrument, control, communication, hydrology, and meteorology. Thus, to obtain the basic knowledge of the technology system calls for a thorough understanding of all the three aspects.

In addition, we also need to understand navigation from the perspective of system construction in that the variety of navigation technologies requires a large amount of corresponding navigation infrastructure construction. For example, the well-known satellite navigation systems (e.g., Beidou, GPS, GLONASS, and Galileo), radio navigation systems (e.g., Loran-C and TACAN), and the more recent developing underwater acoustic navigation systems with long and short baselines all need the support of complex space constellations, land base stations, underwater beacons, and monitoring, computing, and control systems. These unique professional characteristics determine that navigation has a rich connotation and involve a wide range of fields. At the same time, since there are different understandings and definitions of navigation, the division of technology and equipment systems also varies with their own characteristics.

Nevertheless, the core issue of navigation is to determine the motion parameters or to determine the spatiotemporal reference of the moving carrier. This is the focus of most navigation books. Diverse navigation technologies tend to be related to a wide range of fields, such as optoelectronics, underwater acoustics, radar, and communication. Although navigation technologies can be classified into inertial navigation, radio navigation, satellite navigation, underwater acoustic navigation, optical navigation, and environmental feature-matching navigation according to technical means, they are actually quite different from each other in terms of their respective technical fields. For example, celestial navigation and visual navigation are based on optoelectronic technology. The echo sounder, Doppler log, and underwater acoustic positioning system are based on underwater acoustic technology. Navigation radar is an application of radar technology. Satellite navigation and radio navigation are based on wireless communication technology, time-keeping, and measurement technology. Inertial navigation involves more complex technologies associated with precision machinery, optics, materials, mechatronics, and control. When we consider multiple technology methods (e.g., matching navigation based on environmental features, satellite navigation, integrated navigation and all source navigation), the technical fields involved in navigation will be more extensive.

Through years of teaching and research, we have realized that the distinct interdisciplinary characteristics of navigation can be a double-edged sword. It has created ample room for development and innovation of the discipline, but it has also caused difficulty for beginners in this field. Students and junior navigation technicians may often find it difficult to obtain a systematic knowledge or a full picture of this field. Nevertheless, beginners are often encouraged to establish an overall technical framework for navigation before building expertise in a specialized navigation technology. This gives us good reason for writing this book.

To better meet the needs of readers, this book includes the latest development of navigation technology and consists of ten chapters that are divided into three parts: navigation fundamentals, navigation principles, and ship navigation systems.

Chapters 1–3 explain the navigation fundamentals by introducing the concept and role of navigation, the navigation coordinate system, and the concept of positioning-navigation-timing (PNT) parameters.

Chapters 4–9 describe the navigation principles. From the perspective of different navigation parameters and internal technical logic, the various navigation systems are classified into six categories: direct positioning, vector-based attitude determining, velocity measurement, gimbal-based attitude determining, dead reckoning, and integrated navigation. The discussion of the commonalities and differences may help to make it easier for readers to quickly understand and master various principles of navigation systems.

Chapter 10 is about the integrated bridge. From the perspective of navigation equipment system, this chapter introduces the integrated bridge equipment, including electronic charts, navigation equipment, marine hydrological, meteorological equipment, and autopilots. This chapter is also an echo of the last two layers in the functional technical connotation (out of the three layers of navigation) discussed in Chap. 1.

Chapters 2–3 of this book are written by Prof. An Li. Chapter 6 is written by Lecturer Heng Ma. Chapter 7 is written by Lecturer Rongying Wang. Professor Hongwei Bian wrote the other chapters and edited the whole manuscript. During the writing, we received the help of many instructors from the Naval University of Engineering, including Jiangning Xu, Shaofeng Bian, Yongbing Chen, Tao Zhu, Gang Zhou, Miao Wu, Houbu Li, Hao Chen, and Bao Li. Several doctoral students and postgraduate students, including Chaojiang Wen, Wenchao Liu, Haifa Dai, Jun Tang, Yaojin Hu, Zhe Wen, and Li Lin, have conducted extensive data collection and organization work for this book over the past five years. This book also draws on the excellent textbooks and monographs of several renowned scholars in the field of navigation both domestically and internationally in recent years. Finally, we would like to express our special thanks to Cecilia Cao, associate professor of Zhongnan University of Economics and Law, for her careful guidance in English expression.

Navigation technology has developed rapidly to involve a wide range of fields, and this book is a preliminary attempt to establish a unified navigation knowledge system. Due to the limitations of the authors, there will inevitably be shortcomings in the book, and feedback and suggestions from readers and colleagues are greatly appreciated.

Wuhan, China   Hongwei Bian
January 2023

Contents

Chapter 1
Introduction

In new places, people need to know where they are. Where am I? Where are we going? How should I go? These questions are navigation issues. The development of navigation technology comes from human instinct and is the basic requirement of human society.

Navigation is one of the oldest human sciences and technologies. As early as ancient times, human beings had to travel between the jungle, mountains, deserts and oceans to complete all kinds of activities necessary for survival. They had to resort to and create various ancient navigation methods. They either take the distinctive peaks and rivers as symbols, mark various mysterious and strange artificial symbols as references, use the sun, the moon and the mythical stars moving around, or invent ancient machinery such as a south-pointing cart and magnetic compass. With the development of human beings, the expansion of living space and the change in transportation mode, not only people need to navigate but also all kinds of human-made vehicles need to navigate. The continuous development of social needs has been the internal impetus of the development of navigation technology from ancient times until today. If people carefully observe, they will find many navigation technology application services in our life, such as electrical maps, shared bicycle positioning, taxi software and direction finding and positioning technology used in mobile terminals. From these, they will truly feel that the progress of navigation technology not only influences the ability of human control of their own activities precisely but also profoundly affects human society.

To understand the outline and important role of navigation, this section starts with bionavigation and then introduces the development history of navigation technology.

© Science Press 2024
H. Bian et al., *Essentials of Navigation*, https://doi.org/10.1007/978-981-99-5636-4_1

1.1 History of Navigation

1.1.1 Bio-navigation

Like humans, most creatures need to solve navigation problems. During the long evolutionary process, almost all animals have evolved to obtain the necessary navigation capability. Most creatures have the mysterious ability to return to their original place over a long distance or to return to the same place many times according to certain regulations.

There are many reports about the navigation capability of birds, mammals, fish, insects and other animals. For example, the Arctic fox can migrate from the Pacific coast to the Atlantic coast in a few months, which is approximately the same as the journey across Canada from east to west. One kind of shorebird, the spotted-tailed snipe, can fly nearly half the world from its breeding grounds in Alaska to New Zealand. Another North American black-crowned and White-cheeked warbler can fly from the northeastern United States to South America for winter at a very fast speed with the help of trade winds after a 100-h sea flight. Fish also have excellent navigation capabilities. The slender European eel matures and flows down living European rivers for thousands of miles to the Sargassum Sea to lay eggs. After the eel seedlings were born, they migrated back to Europe. Navigation capabilities not only help creatures accomplish incredible long-distance migration but also help them achieve daily predation activities. A dog-kissed bat in Texas, for example, can fly more than 70 km from a cave in search of food. Relative to their size, Saharan desert ants can forage as far as 500 m from their nests. They can accurately find their way home by calculating the number of steps and according to different sunshine patterns and scent clues. This ability is critical to them because they may die from excessive exposure to sunlight.

Why do creatures have these amazing navigational capabilities? Bionavigation has been studied for many years. A large number of examples show that there are still unknown complex navigation biochemical mechanisms in creatures. These methods include the following:

(1) Identify the route by comprehensive analysis of sensory organs. For example, many animals will use a variety of navigation methods synthetically to obtain relevant environmental information from air, water flow, temperature changes, visual markers, odors and other ways to make comprehensive judgments without losing their way. If birds migrate at dusk, they will use the sunset to determine the westward direction; at night, they will navigate by identifying the stars. Pigeons use the sun as a compass to determine where they are flying, calculate the flight distance based on biological clocks and so on.

(2) The instinctive geomagnetic induction compass organism. Many animals have magnetic induction systems that humans may not have. It was found that the heads of animals such as turtles, whales, some birds, some fish and moles have special cells containing magnetic substances. These magnetic substances are

arranged in the direction of the magnetic force line under the influence of the Earth's magnetic field, and the arrangement information is transmitted to the brain for analysis and to issue to control the direction of the animal's movement.

(3) Animals have the ability to visualize the earth's magnetic field. In addition to relying on magnetic cells to induce geomagnetic fields, birds may also visualize magnetic fields with their own X-ray vision system. Some birds have photoreceptors that detect magnetic fields in their eyes. Perhaps in their eyes, the South and the North show different colors.

(4) Orientation is achieved by the perception of sunlight polarization. The direction of polarized light at any point in the sky is perpendicular to the plane composed of the sun, the observer and this point. Therefore, the position of the sun can be determined according to the pattern of sky polarized light. The bee's compound eye is sensitive to polarized light and can detect brightness in different directions in the sky. It has a special directional function. Even if the sun is covered by dark clouds, time can be corrected according to the direction of the sun. Many nocturnal animals, such as cockroaches, can use moonlight to navigate. Therefore, organisms can identify their route by mountains, rivers, coasts or some other visible road signs and can also use olfaction, magnetic fields, starlight and other methods to determine their orientation.

In fact, navigation ability is one of the fundamental abilities of creatures. **It is difficult for a creature to survive without the ability of navigation**. Therefore, it is easy for us to understand the importance of navigation intuitively. In addition, further research reveals the intrinsic organisms of bionavigation, which can help people study and improve new navigation technologies, such as polarized celestial compass. Especially today, with the development of artificial intelligence technology, the importance of bionavigation research is more prominent, and it has become a hotspot of navigation technology research.

1.1.2 Ancient Navigation Technology

Modern humans evolved from Homo sapiens 70,000 years ago. Navigation technology is an essential requirement for human beings of higher animals. Therefore, exploring the history of human navigation can be traced back to early human history, and almost all the world's ancient civilizations have a record of navigation technology. This section mainly introduces the development of ancient navigation technology in China and the Western world and its influence on the development of human history.

Navigation Technology in Ancient China

China has a long history innavigation technology research. Historical records show that the earliest navigational equipment in China was the Southward Pointing Cart

Fig. 1.1 The southward
pointing cart, photograph by
Bian H. W.

more than 4000 years ago (Fig. 1.1). From today's technical point of view, the southward pointing cart is a navigation device that maintains its original azimuth accurately, and it is also one of the representative inventions of ancient Chinese machinery. According to the legend, it was invented by Yellow Emperor (Huang-di), who fought against his powerful opponent (Chi-you), a mythological warrior, in Zhuo-lu (a place in Hebei province) but encountered heavy fog. It was by using the southward pointing cart to identify directions of the troop that Chi-you was seriously defeated. From this legend, we can also see that as a navigation device, the southward pointing cart was closely linked with military applications at its first invention in ancient times.

In the Xia Dynasty more than 4000 years ago, Chinese ancestors had learned to navigate by using geographical targets to reach their destinations. "Shangshu-Yugong" (an ancient book) records that the minority nationalities in eastern Liaoning in the Xia Dynasty (2070 BC–1600 BC) usually sailed in the Bohai Sea, passing through Jieshi Mountain on the right and entering the Yellow River estuary to the capital of ancient China. Historical records show that before the fourth century BC, ancient Chinese were able to sail freely in all the surrounding seas.

The first record of the compass in China was found in the Spring and Autumn Period and the Warring States Period (770 BC–221 BC). Around the first century BC, Chinese wizards used a magnetite spoon in the shape of the Big Dipper, which is placed on a smooth copper plane, to indicate the north (Fig. 1.2) [1]. At least 1500 years ago, the compass was used for navigation, enabling people to travel far from the coast into the ocean. Shen Kuo (1031 AD–1095 AD), a famous scientist and litterateur in the Northern Song Dynasty (960 AD–1127 AD), first described the discovery of the magnetic declination angle in his book ("Meng xi Bi tan"). By approximately 1090 AD, it was recorded that Chinese sailors of naval fleets identify the direction by various methods, e.g., by stars at night, by sun at day, and by using the magnetic compass on cloudy and rainy days. In the Song and Yuan

Fig. 1.2 Sinan invented in the Han Dynasty, photograph by Bian H. W.

Dynasties (960 AD–1368 AD), China's navigation technology was well developed, and maritime trade was extremely prosperous. Later, the magnetic compass was introduced to Arab and Europe and played a great role in human beings. Therefore, Joseph Needham called the Chinese magnetic compass in the Middle Ages a great invention.

China is one of the earliest countries in the world in the development of astronomy. Ancient Chinese knew Polaris (called in Chinese Beichen, the star of north) very early, so they used it to distinguish direction. As far back as 2000 years ago, ships carried various goods across the sea and traded with Japan, southeast Asia and other neighboring countries. At that time, celestial methods were used to navigate the ocean. Faxian (334 AD–420 AD), a Buddhist monk of the Eastern Jin Dynasty (317 AD–420 AD), wrote when he visited India by boat, "The sea has no end, it can't tell the East and West, sail forward only rely on observing the sun, moon and stars." By the Song Dynasty (960 AD–1279 AD), the celestial navigation method had further developed. Celestial positioning technology developed greatly in the Yuan and Ming Dynasties (1271 AD–1644 AD). However, if you sail in the sea only know the south and north directions but do not know the specific location, you will still lose your course, and it is difficult to reach your destination successfully. With the development of navigation, China has created a celestial navigation technology called "lead star technology", which uses the lead star board to determine the ship's azimuth in the sea. According to the vertical height and the length of the rope measured by the tracer board, the polar star altitude angle is converted to approximately determine the local geographic latitude.

As early as the early thirteenth century, China had the earliest charts of the South China Sea and its islands. However, the earliest chart circulated to this day is Zheng

Fig. 1.3 Ancient Chinese chart

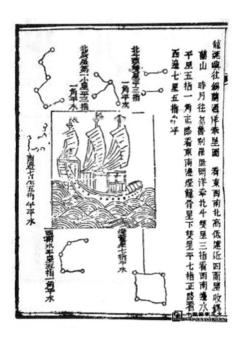

He's chart of the Ming Dynasty in 1430 AD (3). Zheng He (1371 AD–1433 AD), a famous navigator in the Ming Dynasty (1368 AD–1644 AD), led the fleet to the Indian Ocean seven times [2]. The navigational technology adopted by Zheng He was the most advanced navigational technology at that time, based on the knowledge of marine science and nautical charts, using navigational instruments such as magnetic compass, log and bathymeter to ensure the navigation route of ships according to charts and guidebooks. According to records, Zheng He's fleet has crossed the equator and reached more than 30 countries and regions. In the daytime, a 24/48 azimuth magnetic compass is used for navigation, and in the evening, stars and a magnetic compass are used. The integrated application of navigation technology made ancient Chinese navigation technology lead the world before the fifteenth century (Fig. 1.3).

Navigation Technology in the Ancient World

The earliest records of ancient navigation technology in the West can be traced back to the Phoenician and Greek periods in BC. They have mastered how to navigate the Mediterranean at the same latitude as east and west by observing the altitude of the Polaris. The ancient European navigators used to regard "wind direction" as "direction". Early Nordic pirates used geographic navigation in northern Europe's North Sea and Baltic Sea waters. At that time, the captain was very familiar with the sea and natural things. He also used information such as birds, fish, currents, driftwood, seaweed, water color, ice reflections, clouds, wind and other information to navigate. In the ninth century, the famous Nordic navigator Fletcher even guided him to Iceland by releasing a raven on board.

Charts are essential for navigation. Ancient Egypt, Greece and Arabs contributed to the development of charts, put forward some basic theories and drawing methods, and drew corresponding early charts. Among them, the famous representative was Claudius Ptolemaeus of Ancient Greece (90–168 AD). He published an important book "Geography Guide" on the basis of celestial and geodetic research more than 1800 years ago. Using the latitude and longitude network established by Hipparchus, the circumference is divided into 360 parts, each location is marked with latitude and longitude coordinates, and 8100 locations in Europe, Asia and Africa are listed. The 27 world maps and 26 regional maps drawn are collectively called Ptolemaeus Maps. In the thirteenth century, with the emergence of the Portolan chart, the chart was separated from the map, forming an important independent branch of the map, which prepared the conditions for the great development of navigation and the great discovery of geography, promoted the Renaissance, and initiated the road of the whole revival of cartography in the late Middle Ages. This chart regards the ocean as the main object of expression and only shows the coastal narrow and long area on land. All the objects that are useful for navigation are markedly represented. The chart of the bay maintains the correct position and orientation of the targets. The contours of the coast and island are similar to those of the corresponding scale of modern charts. It is of pioneering significance in the design and application of symbols and colors. The Renaissance in the 15th to sixteenth centuries gave birth to great geographical discoveries. To meet the needs of navigation and exploration, frequent navigational activities required that charts should contain more content, cover more extensive areas, have higher accuracy and be more convenient for navigation. As a result, Mercator charts were created, which is another milestone after Ptolemaeus charts and has been used up to now.

In terms of geomagnetic navigation, the Arab Empire learned to use the compass from the Tang Dynasty (618 AD–907 AD) in its heyday, which greatly stimulated the great development of Arab navigation. In the twelfth century, the magnetic needle was introduced from Arabia to Europe. In 1190 AD, Italian pilots began to float iron needles in a bowl of water, magnetize them with a natural magnet, and check the direction of the needle. By approximately 1250 AD, it had developed into a marine magnetic compass. At the beginning of the fourteenth century, Joya, an Italian, used paper to make a direction dial and a magnetic needle to drive together. This is a leap in the development of magnetic compasses. In the sixteenth century, the Italian Cardan designed the gimballed ring, which had been invented and used by the Chinese as early as the Han Dynasty (202 BC–220 AD), but Cardan used it to control the magnetic compass to maintain its level during the ship's rolling. The Europeans then converted the original compass into a simple marine compass and mapped the Mediterranean Sea. Maritime magnetic compass is widely used in navigation. In the eighteenth century, when the steam engine was invented, steel used in ships and the usage of a large amount of steel produced a strong magnetic field. The magnetic field produced by steel made huge and regular errors in the traditional compass. This problem aroused the attention of French scientist Poisson, British astronomer Erie,

Russian navigator Ivkruzgere and others and made continuous research on the marine magnetic compass. Research and improvement eventually led to the formation of a simple compass that played a great role in history and is still used today with complete instructions, calibration, lighting, and observation system magnetic compass.

In celestial navigation, in 1637, French scientist Reńe Descartes published "Geometry", which created a plane rectangular coordinate system to describe points in space, successfully created analytical geometry and laid a mathematical founda tion for modern navigation and positioning theory. In 1730, the navigational sextant was invented to measure the latitude of the observer by observing the altitude of the Polaris. In 1767, Harrison, a British clockmaker, invented the chronometer. By combining it with the sextant, the longitude of the observer was measured. In 1837, an American captain discovered the contour method and used it to determine the latitude and longitude of the vehicle position. In 1875, the French navigator St. Hille proposed the principle of the altitude difference method, which laid a theoretical and practical foundation for celestial navigation.

Around the middle of the nineteenth century, sextant, chronometer, magnetic compass, bathymetric ballast and log have been widely used in the field of navigation, which has led to the development of marine surveying and mapping technology. Most of the world's coastlines have been surveyed, and charts have been drawn accordingly. The safety of navigation at sea is basically guaranteed.

1.1.3   Modern Navigation Technology

Since the beginning of the twentieth century, many significant technological break-throughs have been made in the field of navigation. A series of new navigation systems, such as inertial navigation, radio navigation and celestial navigation, have emerged, which have made great progress in many fields, such as space, aviation, maritime and underwater exploration.

(1) Inertial navigation. The first gyrocompass appeared in 1908, followed by the gyro horizontal instrument developed in the 1920s. In 1942, Germans invented the V-1 rocket and the V-2 rocket. V-2 inertial guidance system is considered the initial form of the inertial navigation system. Inertial navigation systems (INSs) have attracted worldwide attention since their inception because they can provide the vehicle position and orientation in three-dimensional space in all weather, autonomously and continuously in any environment. At present, it has been extended to the fields of aerospace, aviation and maritime applications. It has also been further used in the fields of mineral exploration, oil exploitation, geodetic mapping, ocean surveys, earthquake prediction, subsea lifesaving, submarine cable laying and positioning.

(2) Radio navigation. In fact, the operation process of a radio navigation system is to receive, process and convert the geometric parameters of the receiving point to the coordinates of the navigation station by utilizing the effective resources of

radio signals and then to measure the navigation parameters such as the direction, distance and distance difference between the moving object and the navigation station based on the electrical parameters of radio navigation signals. It has many kinds, such as radio direction finder, Loran (remote radio navigation system), TACAN (tactical air navigation system), Decca and satellite navigation systems, such as GPS, GLONASS, Beidou, Galileo and so on. In addition, there are many common navigation devices based on radio technology, such as navigation radar, meteorological fax machines, and AIS systems.

(3) Celestial navigation. Before World War II, celestial navigation was the main means of navigation on the sea. Almost all ships are equipped with various astronomical calendars, chronometers and sextants for celestial navigation. As general navigation equipment, they are still used today. After World War II, celestial navigation periscope equipment combined with sextant and periscope appeared. With the development of technology, inertial stabilization platforms have become an integral part of celestial navigation systems, providing a high-precision horizontal reference for celestial navigation. With its own character-istics, celestial navigation has been widely used in many fields, such as space, aviation, and maritime.

(4) Acoustic navigation. The earliest underwater acoustic navigation technology can be traced back to the birth of sonar in 1917. In 1925, the United States developed the world's first acoustic sounder, and since then, the echo sounding method has replaced the traditional hammer method. Acoustic velocity measure-ment equipment mainly includes Doppler logs and acoustic correlation logs. The underwater acoustic navigation system also includes acoustic positioning tech-nology and terrain matching navigation. The acoustic positioning system can be divided into a long baseline system (LBL), a short baseline system (SBL) and an ultrashort baseline system (USBL) according to the length of the base-line between bacons. Currently, underwater acoustic navigation technology has become a global research hotspot.

In addition to the above navigation technologies, matching navigation based on environmental characteristics, landmark navigation, visual navigation, sound navi-gation, integrated navigation, all-source navigation and other technologies have also been fully developed and applied. From the development of navigation, we can see that navigation technology has a profound impact on social development. It is not only the basic need of human beings but also the comprehensive embodiment of national power. At the same time, the navigation system has a variety of types, a wide range of technical fields, prominent comprehensive application, and a strong driving force for development.

1.2   Concept of Navigation

The concept of navigation originated from navigation on the sea, which is generally understood as the process of guiding the vehicle from the origin to the destination. With the development of science and technology, a variety of means of delivery, such as aircraft, missiles, rockets, giant oil tankers, artificial satellites spacecraft and so on, have appeared in succession, greatly expanding the concept of "navigation". Modern navigation puts forward higher requirements for navigation instruments or equipment, which are no longer simply to ensure the safety of vehicle navigation but also to provide much navigation information, such as vehicle speed, course, water depth, attitude (pitch, roll), position (longitude, latitude), time, and so on. Gradually, the ship's course and track can be controlled, and even the ship's dynamic positioning can be realized.

The traditional definition of "the process of guiding the vehicle from the origin to the destination" is also different from definitions such as power technology, digital network technology or photoelectric technology. It is described as a process, a function or an application, rather than a specific technical form. In fact, navigation has a few limitations on technology. As long as the technology can accomplish the above purposes, it can be adopted as navigation technology. Therefore, there are many different navigation technologies.

What is the connotation of the concept of "navigation"? What are the core elements of its technology? In recent years, people have tended to divide navigation technology into several different levels. From our points of view, navigation technology solves the problems for human beings in three different categories, namely:

(1)   Measuring the motion parameters of the vehicle itself;
(2)   Measuring the environmental information of the vehicle;
(3)   Make route planning and guide the carrier to the destination.

Three aspects of navigation technology are relatively independent and closely related, which can basically display the panorama of navigation technology.

1.2.1   Measuring the Motion Parameters of the Vehicle

This is the key function of the navigation system, which is the general understanding of the function of navigation. We can consider this as knowing the information of the vehicle "itself". The motion parameters of the vehicle to be determined generally refer to the coordinate information of the vehicle itself in four-dimensional space–time. These motion parameters can be divided into three categories: translational parameters, rotational parameters and time parameters.

(1)   **Translational parameters of the vehicle**: If the vehicle is regarded as a mass point, the navigation system should be able to provide information about the position, velocity and acceleration of the vehicle in three-dimensional space.

(2) **Rotation parameters of the vehicle**: In reality, the vehicle is similar to a rigid body more than a mass point. In this situation, the navigation system should be able to provide information such as the attitude angle and the angular velocity of the rigid body in three-dimensional space.

(3) **Time parameters of the vehicle**: The navigation system should also provide accurate time and frequency references.

Generally, when we need to determine all or part of the above information of the vehicle, we should resort to navigation technology to solve the problem.

There are many kinds of navigation systems and equipment. According to the different information provided, the navigation systems or equipment can be divided into different kinds, such as positioning systems, direction finding equipment, velocity measuring equipment, attitude measuring equipment, timing equipment, and stabilization systems. According to the different technical methods used, navigation technologies can be divided into different kinds, such as inertial navigation, radio navigation, satellite navigation, acoustic navigation, optical navigation and matching navigation. These technologies are quite different from each other. To some degree, a technology that can provide the motion parameters of the vehicle in space and time can be categorized in the navigation research domain.

1.2.2 Measuring the Environmental Information of the Vehicle

It is not enough for all kinds of vehicle navigation to only determine the motion information of the vehicle itself but also to require the environmental information of the vehicle; otherwise, it will lead to various accidents. Examples include traffic accidents involving vehicles, collisions and grounding of ships, etc. Generally, to navigate also has to know "the other".

According to the different environments of the vehicle, the navigation system is divided into four different geographic fields: aerospace, aviation, maritime, land and so on. Accordingly, there are four different application fields of navigation technology. For example, the navigation of ships belongs to marine navigation, the navigation of tanks and self-propelled guns belongs to land navigation, the navigation of aircraft and missiles belongs to aviation navigation, and the navigation of rockets and satellites belongs to aerospace navigation.

Navigation-related environment information is abundant and usually includes the following:

(1) **Meteorological information**: air temperature, air pressure, humidity, wind direction, wind speed, airflow, cloud, fog, rain, snow, visibility, weather, etc.

(2) **Hydrology information**, including water depth, water temperature, salinity, density, surge, ocean currents, tides, sea conditions, etc.

(3) **Geographic information**, including topography, islands, reefs, obstacles, etc.

(4) **Dynamic Environment Information**: This includes the information of surrounding vehicles, such as ships, flying objects and underwater vehicles, which have an impact on navigation safety.

The above information will affect vehicle motion control and navigation safety, involving a variety of different functions, different working principles, different technical fields of instruments and equipment, such as meteorological fax machines, anemometers, meteorological maps, bathymeters, hydrology devices, navigation radar, AIS systems and other radio navigation devices, and electronic charts. Sometimes these devices are categorized as navigational equipment. Sometimes they are also categorized as specialized meteorological equipment, hydrological equipment, avionics equipment and so on.

1.2.3 Route Planning and Guiding the Vehicle to the Destination

Navigation systems also need to analyze and determine the motion control scheme of the vehicle according to various environmental factors. The first two aspects of navigation actually solve the problem of where the vehicle itself is and what about the surrounding environment? However, there are no solutions to how to maneuver, how to drive, and how to achieve the goal according to the sailing intention. This problem is one of the ultimate demands of all kinds of navigation problems.

It includes decision-making, judging, monitoring and control. Commonly speaking, it is "routeplanning and control". It can also be divided into two separate categories. In this book, considering that both of them are closely related, they are classified as one level.

(1) **Decision-making**, including route planning, safe collision avoidance, track control, situation monitoring and maneuvering calculation; taking ship navigation as an example, it is the most important task for navigation departments to ensure a safe voyage.
(2) **Vehicle control** includes specific control of the actual course, track, speed and attitude of the vehicle, such as ship navigation control and aircraft flight control. It is closely related to the overall motion characteristics and control characteristics of the vehicle. According to the different vehicles, it can be divided into carrier navigation and weapon guidance. In the field of vehicle motion control, there are not only ways to control ships, aircraft and other vehicles to reach the planned destination but also ways to control missiles, artillery shells, torpedoes and other weapons to hit their target.

The decision-making function endows modern navigation equipment with distinct technical attributes of information-related, intelligence and networking and possesses typical characteristics of information equipment. The navigational equipment includes avoyage management system (VMS), voyage data recorder (VDR), electrical chart of display and information system (ECDIS) and so on. The vehicle control function embodies the control, and the manipulation system is a typical control system. Mainly includes autopilot, fin stabilizer, dynamic positioning system (DPS) and so on.

1.3 Navigation Related Professional Fields

Navigation is interdisciplinary. Different navigation technologies often belong to different professional fields. This section briefly introduces the concepts and research fields closely related to navigation. From the comparisons below, we can deepen our understanding of the concept of navigation and its professional characteristics.

1.3.1 Navigation Versus Guidance

Navigation and guidance are closely related. The differences between navigation and guidance are mainly embodied in the following aspects:

(1) Different working methods. Navigation is a device that manually operates and guides the vehicle to reach the destination according to the predetermined route. The navigation system can be regarded as a navigation parameter measurement device, and the main task is to output the navigation information. Guidance is based on the measured navigation parameters, through the control system solution, direct control of the object according to the predetermined route to the destination, which is closely related to the autopilot.
(2) The device accuracy of inertial guidance systems is usually lower than that of inertial navigation systems. This is due to the short running time of missiles and rockets, which can ensure the design accuracy of the system in a short time.
(3) The vibration of missile and rocket launching is much more violent than that of aircraft and ship launching, so the strength, vibration resistance and reliability of the guidance system are more stringent than those of the navigation system.

1.3.2 Navigation Versus Geodesy

Geodesy, also known as surveying or mapping, is a discipline to study the theory and calculation of geometric positions of the earth surface, earth shape, earth gravitational field, natural objects and artificial facilities on the surface, and the preparation of various scale maps. The research object of geodesics is geospatial information such as the form, position, gravity and other distribution of the earth, so geodesics can be considered a branch of geoscience. In recent years, the research objects of mapping have also expanded from the Earth's surface to outer space and the Earth's internal structure.

Navigation and geodesyare closely related. The most direct examples are maps, charts and a large amount of geophysical field information used for navigation, which all depend on surveying and mapping. Navigation is used for dynamic measurement of vehicle motion parameters, and geodesy is used for long-term static measurement of background field information. They have common mathematical foundations in the description of various coordinate systems of the earth, map projection and other aspects. Surveying and mapping can adopt static measurement and postprocessing methods. The measurement means, equipment and data processing methods can often provide more accurate references, which information can also be applied to navigation equipment for dynamic measurement, such as calibration and verification.

1.3.3 Navigation Versus Detection

Detection systems include radar, sonar and other target detection systems. Navigation and detection are also closely related. Both of them have similar theories in the calculation of the moving parameters of the vehicle, but the detection system is mainly used to calculate the motion parameters of the target, while the navigation system is mainly used to calculate the motion parameters of the vehicle itself.

Some navigation systems and detection systems have similar technical means. Some navigation means based on external measurements are closely related to the detection method. Detection systems often detect unknown targets, while navigation systems mainly detect known targets. For example, positioning based on navigation radar actually locates the vehicle itself by detecting the range and azimuth of the fixed target. In terms of technical means, they all adopt the same radar technology. Similarly, sonar can be used to measure the known underwater acoustic beacon to achieve the self-localization of the underwater vehicle.

In practical applications, they are often used together. It is detection technology when detecting the motion parameters of the target. When detecting the motion parameters of its own vehicle, it is classified as navigation technology. The detection source can be placed outside or inside the vehicle. For example, the spacecraft itself has an independent navigation system, while there are many detection systems on the ground to track and monitor the motion of the spacecraft. When the external detection system can obtain more accurate vehicle motion parameters, it can be used as the external benchmark of the vehicle's own navigation system to test and evaluate the vehicle navigation system and motion control system. For example, satellite navigation systems precisely use ground monitoring stations to monitor satellite motion and control satellite entry into orbit precisely. In the process of landing an aircraft for the aircraft carrier, it is usually accomplished by the joint cooperation of the detection and navigation system on the surface of the ship and the airborne navigation system. In the process of ship maneuvering, target positioning and maneuvering control also need to integrate the information of the detection system and navigation system.

1.3.4 Navigation Versus Communication

Navigation and communication are closely related. Navigation uses many communication technologies. The technical field of communication is very wide. According to the mode of transmission, it involves wireless communication and wired communication. According to the technical classification, it involves photoelectric communication, underwater acoustic communication, radio communication, radar communication and so on. Communication systems mainly transmit information, while navigation systems extract navigation-related information about spatial distance and azimuth during the transmission of information.

Navigation systems need to transmit corresponding navigation information by means of communication technology, and different navigation reference stations (or satellites) need to confirm and distinguish each other by means of communication coding technology. For example, satellite ephemeris in satellite navigation systems needs to be transmitted by means of microwave communication technology, and the Loran C radio navigation system can realize the transmission of station chain information. Similarly, the underwater acoustic navigation system also needs some coding design for the transmitted acoustic signals, and its basic equipment has many similarities with many underwater acoustic communication systems. The commonality between them is that both the sender and the receiver are needed to complete the information transmission by means of waves. The difference between them is that the purpose of communication systems is to transmit information, which has wider applications, while that of navigation systems is to acquire basic motion elements such as the position, speed and heading of vehicles.

In some application areas, communication and navigation also appear to be directly integrated. For example, the relative navigation system of a formation adopts the communication link of a fleet to realize the solution and sharing of navigation information, which can help to enhance the navigation ability of each combat unit. The short message function of the Beidou system is a kind of communication function that can monitor multiple users by means of communication.

1.3.5 Navigation Versus Maritime

Navigation and maritime are closely related. In the development of navigation technology, in a long historical period, maritime demand drove the development of navigation technology. Compared with land navigation, marine navigation can use less

external information and therefore is more difficult. In addition to marine navigation, in the twentieth century, humans gradually developed aerospace navigation and
aviation navigation.

Marine technology has a wide range. According to the definition of marine, it
involves oceanography, hydrology, meteorology, shipbuilding technology, power
technology, material technology and so on. Therefore, it has a broader professional
field. "Maritime" often corresponds to "Aviation" and "Aerospace". According to
the definition, aviation is a comprehensive technical discipline that applies various
methods to accomplish in-atmosphere flight. Space flight refers to the space science
and technology related to the research and exploration of outer space. Navigation
technology is applied not only to marine navigation but also to aviation, aerospace
and land transportation. Therefore, there is an overlap between these two domains.

1.3.6 Navigation Versus Measurement and Control

Measure and Control Technology and Instrument (MCTI) is a technology used
to study the acquisition and processing of information and the control of related
elements. It is a high-tech intensive comprehensive subject formed by the inter-
penetration of electronics, optics, precision machinery, computers, information and
control technology.

There is a close relationship between navigation and MCTI. A large number of
sensors and measurement and control technologies are used in navigation systems
to obtain vehicle navigation information and environmental information. A large
amount of navigation equipment, e.g., logs, bathymeters, and wind direction finders,
is a typical MCTI, which adopts the various design architectures of measurement and
control equipment. Not only do the devices in the navigation system use much sensor
technology, but the navigation system as a whole can also be regarded as a sensor
system by the high-lever user system. At the same time, many control technologies
are involved in the navigation control of vehicles or the control of inertial stabilization
platforms.

Appendix 1.1: Questions

1. Can you understand the importance of navigation technology from the perspectives of biological instinct, personal needs, social development, national security
   and human progress?
2. According to the definition of navigation, we analyze the different technical levels
   in the broad concept of navigation.
3. List the navigation parameters of the carrier. How can these parameters be
   classified? What navigation devices can provide these parameters?

4. How can the environmental information and parameters related to navigation be classified in detail? What navigation devices are involved?
5. Please classify the navigation devices from the technical and functional perspectives, give an example, and analyze the differences and connections between the two classification methods.
6. Please analyze the relation and differences between navigation and related techniques from the perspective of universal connexion.
7. Why is navigation an important and basic part of the modern information war?
8. Thinking about the experience and lessons that should be summarized from the history of Chinese Marine civilization?

Appendix 1.2: The Origin, Development and Reflection of Chinese Marine Culture

With a vast territory, China is both a continental country and a maritime country, and such a geographical characteristic has determined the diversity of Chinese civilization since ancient times. It is generally believed that Chinese civilization originated in the Yellow River basin, and research has confirmed that the Chinese maritime civilization, a branch of the indigenous civilization, started to develop at the same time as the Chinese agricultural and nomadic civilization did.

Such a unique maritime culture can be traced back to the Chinese Dongyi (东夷) and Baiyue (百越) people, and communication has long been established between the coastal areas of Zhejiang, Fujian, and Taiwan and those of Southeast Asia, Australia and even America. In fact, the coastal areas of China have always had a strong navigation tradition and influence. The long-term advanced navigation technology in ancient China has also witnessed this influence. Zheng He's voyages to the western oceans in the 1400s was the pinnacle of ancient Chinese navigation, and together with Zheng Chenggong's recovery of Taiwan in 1661, the event revealed the leading position of ancient China in the field of navigation at that time. Therefore, it is safe to say that the Chinese nation has a long and splendid maritime culture and a maritime spirit of having the courage to explore and advocate harmony. A better understanding of the Chinese maritime traditions can make it easier for readers to make connections between the past and the present.

Moreover, human civilization presents a sense of diversity thanks to different natural conditions, ways of life and social history. Hence, exchanges and mutual learning between different civilizations are of great significance in solving the problems in our own development and at the same time promoting the common progress of human civilization. People should not only cherish their own culture, but also respect and appreciate the cultures of other nations. In the face of different civilizations, we should learn from each other with an open mind, rather than making subjective assumptions, or jumping to unsupported conclusions. The world calls on us to walk our own path of modernization, and that can only be achieved based on our own civilizations and actual national conditions and with the spirit of openness, inclusiveness, equality, win–win, and innovation.

Chapter 2
Navigation Coordinate System

Why does navigation study coordinate systems? We know that describing object motion is relative to a certain reference frame. The task of the navigation system is to determine the motion parameters of the carrier, that is, to determine the series of navigation parameters such as the carrier position, speed, attitude and its change rate in a certain coordinate system. In other words, the navigation system itself is a reference coordinate system for the carrier, which is often known as the spatiotemporal reference. Therefore, when studying the navigation problem, it comes first to determine the coordinate system, which is not only the mathematical basis for defining navigation parameters and navigation solutions but also the basic problem of all navigation technologies and navigation systems. It is very important to describe navigation parameters and solve navigation problems. This chapter will highlight the relevant knowledge about the navigation coordinate system.

2.1 Main Coordinate Systems

2.1.1 Types of Coordinate Systems

Cartesian Coordinate System in Three-Dimensional Space

The coordinate system was first put forward by Descartes in 1637, namely, the rectangular coordinate system, which mainly includes a two-dimensional plane rectangular coordinate system and a three-dimensional space rectangular coordinate system.

The coordinate system has three basic elements: origin, axis, and unit. To select and determine a coordinate system, it is usually necessary to determine (1) two poles, namely, the base pole and the second pole; (2) two planes, namely, the base plane and the second plane; and (3) three axes, namely, the base axis, the second axis and the third axis. The base pole (line) is the symmetrical axis of the coordinate system,

© Science Press 2024

H. Bian et al., *Essentials of Navigation*, https://doi.org/10.1007/978-981-99-5636-4_2

Fig. 2.1 Cartesian
coordinate system

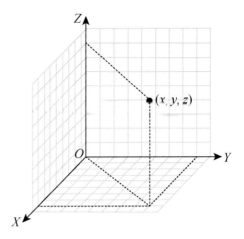

such as the rotation axis of the earth. The base plane is a plane perpendicular to
the base pole, such as the equatorial plane. The second plane is perpendicular to
the base plane and contains the base pole (line), such as the Greenwich meridian
plane. The second pole (line) is the intersection line between the base plane and
the second plane. The third axis is orthogonal to the base axis and the second axis,
forming a right-handed rectangular coordinate system. Once the above elements are
determined, a coordinate system is basically determined.

   All points in space can be projected to three coordinate axes, and their coordinates
$\{x, y, z\}$ can be obtained (Fig. 2.1). Three-dimensional space rectangular coordinate
systems are widely used in the navigation field. The inertial coordinate system,
geographic coordinate system, body coordinate system and navigation coordinate
system are all three-dimensional space rectangular coordinate systems.

Spherical Coordinate System

Points in a two-dimensional plane can be described in Cartesian coordinates (X, Y),
which can also be expressed by polar coordinates (ρ, θ) . Similarly, points in three-
dimensional rectangular coordinate systems (x, y, z) can also be described by the
spherical coordinate system (r, α, δ) (as shown in Fig. 2.2), where r is the length of the
vector (OM),α is longitude, and δ is latitude ($0 \leq r < \infty, 0 < \alpha < 2\pi,\ 0 \leq \delta \leq \pi$).

   Spherical coordinate systems are widely used in navigation. Usually, the navigation positioning is on the surface of the earth, which is a sphere. Because the earth is
not a regular sphere, ellipsoidal coordinates are often used in practice. The ellipsoid
coordinate system can be regarded as a general form of the spherical coordinate
system. Geodetic coordinates and celestial coordinates commonly used are spherical
coordinates. Longitude, latitude and altitude are defined in a spherical coordinate
system, and their common symbols are (λ, φ, L).

Fig. 2.2 Spherical
coordinate system

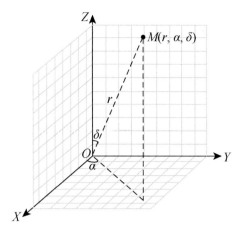

Other Coordinate Systems

Cylindrical coordinate systems and conical coordinate systems are also used in navigation; for example, cylindrical coordinate systems are often used in navigation mapping, such as Mercator projections and Gauss projections.

2.1.2 *Common Navigation Coordinate System*

Inertial Coordinate System

An inertial coordinate system is a coordinate system describing inertial space. The so-called inertial space refers to the space in which an object can keep stationary or uniform linear motion when it is not subjected to force or the resultant force is zero.

The inertial system is a very important concept in mechanics. In the inertial coordinate system, the relationship between force and motion is described by Newton's law. With the development of physics, Einstein created a new space–time theory in the twentieth century, which developed the concept of Newton's inertial system.

The book follows Newton's idea of an inertial system. To establish an inertial coordinate system, we must find a reference object that is stationary or moving uniformly in inertial space. However, according to the principle of gravitation, there is no object absolutely under zero force in the universe, so the ideal inertial reference system cannot be found in reality. On the other hand, in practical applications, it is not necessary to find an absolutely accurate inertial reference system but only to find an approximate inertial reference system. Thus, the approximation depends on the needs of the requirement.

According to different approximations, many different inertial systems can be obtained. For objects other than the sun in the solar system, the Sun-center (heliocentric) coordinate system is very similar to the inertial system, and for objects other than the Earth in the Earth-Moon system, the centroid coordinate system of

Fig. 2.3 Sun-center inertial
coordinate system

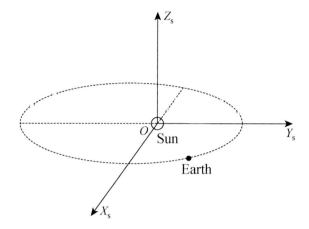

the Earth is very similar to the inertial system. The following are several inertial systems commonly used in navigation. The main difference between them lies in the selection of the origin and the definition of the axis.

(1) Sun-center inertial coordinate system (SCI system)

The origin of the SCI system is at the center of the Sun's sphere. The orbital plane of the Z_s axis is vertical to the Earth's revolution, and the X_s and Y_s axes are in the right-hand coordinate system in the plane of the Earth's orbit of revolution (Fig. 2.3), which is ecliptic. The SCI is suitable for the navigation and positioning of interstellar vehicles. The angular velocity of the sun's rotation around the galactic center is approximately 0.001 angular seconds/year, and the centripetal acceleration of the sun's movement around the galactic center is approximately 2.4×10^{-10} m/s^2. The angular velocity and acceleration of the sun's motion around the Galaxy are far less than the minimum that human instruments can measure at present. Therefore, the use of SCI ignores the movement of the sun in the Galaxy and has sufficient accuracy.

(2) Earth-center inertial coordinate system (ECI system)

The coordinate origin of the ECI system is taken at the geocentric center of the earth. The Z_i axis is along the Earth's axis of rotation, and the X_i axis and Y_i axis are in the equatorial plane of the Earth. The X_i axis, Y_i axis, and Z_i axis form a right-handed coordinate system. The X_i axis usually points to the intersection of the equatorial plane of the earth and the ecliptic plane of the sun, which is called the equinox (spring equinox and autumn equinox) and does not change with the rotation of the earth (Fig. 2.4). The ECI is suitable for studying the navigation and positioning of vehicles near the Earth's surface. Near-Earth navigation mostly uses ECI as an inertial coordinate system. For example, the motion of an Earth satellite can be calculated in the ECI system.

This approximate inertial reference system ignores the acceleration caused by the gravitational pull of the sun, moon and other stars on the earth. The acceleration of the Earth's common centripetal acceleration caused by the sun's gravity is

Fig. 2.4 Geocentric inertial
coordinate system

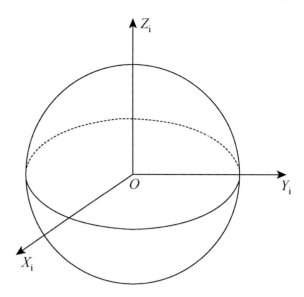

approximately 6.05×10^{-3} m/s^2. The centripetal acceleration of the earth caused by the moon's gravitation is approximately 3.4×10^{-5} m/s^2. The acceleration caused by the gravitation of other stars is two orders of magnitude smaller than that of both above, so the acceleration of the origin of the geocentric inertial coordinate system is approximately 6×10^{-6} m/s^2. This magnitude can be sensitized by current accelerometers, so in some applications of inertial technology, the ECI is not accurate enough. However, if we only care about the motion of objects relative to the earth, the acceleration caused by the gravitation of the sun and other stars to the earth and the near-earth objects is approximately 10^{-6} m/s^2. The magnitude can be neglected.

(3) Celestial coordinate system (C system)

The celestial coordinate system uses spherical coordinates to describe the position of celestial objects (stars, planets, etc.). The position of celestial objects in the celestial coordinate system only indicates the direction of celestial objects without considering the distance of celestial objects. The celestial sphere rotates around the Earth's axis from east to west. The intersection of the axis of rotation and the celestial sphere in the northern (southern) hemisphere is called the northern (southern) celestial pole, namely, $P_N (P_S)$. The northern and southern ecliptic poles are represented by $\Pi (\Pi')$, respectively. The angle between the ecliptic plane and the celestial equatorial plane is called the ecliptic obliquity ε, with an average value of approximately $23.5°$ (Fig. 2.5).

The coordinate origin of the Ecliptic coordinate system is taken at the geocentric center of the earth. The Z_i axis is along the line of $\Pi (\Pi')$, and the X_i axis and Y_i axis are in the ecliptic plane of the earth. The X_i axis usually points to the spring equinox, and the X_i axis, Y_i axis, and Z_i axis form a right-handed coordinate system (Fig. 2.6). Based on the coordination, the ecliptic longitude and latitude are defined.

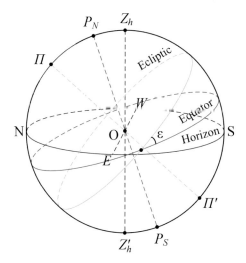

Fig. 2.5   Points and circles on the celestial sphere

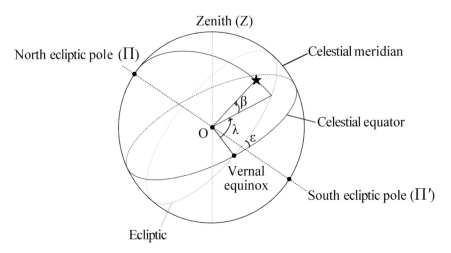

Fig. 2.6   Ecliptic coordinate system

Earth-Center Earth-Fixed Coordinate System (ECEF System)

The ECEF system is also known as the Earth coordinate system (*e* system). The origin of the coordinate system in the center of the earth, because the earth is always rotating, the earth coordinate system and the earth are fixed, rotating with the earth polar axis Z_e. The X_e axis is at the intersection of the equatorial plane and the primary meridian plane, and the *Ye* axis is also in the equatorial plane. The three axes form right-hand rectangular coordinates (Fig. 2.7).

Fig. 2.7 ECEF system and geographic coordinate system (E-N-U)

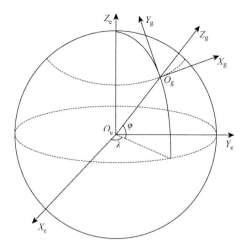

Considering the autorotation and revolution of the earth, the rotation angular velocity of the e system relative to the inertial reference system is

$$\omega_{ic} \approx 15.0411(°/h) \approx 7.2921 \times 10^{-5} \text{ rad/s} \tag{2.1}$$

Earth Surface Coordinate System

(1) Geographic coordinate system (g system)

The g system is the most commonly used coordinate system (Fig. 2.8). The origin of the g system is the observer or the center of the vehicle. The axis X_g can point east in the local horizontal plane. The axis Y_g then points north along the local meridian circle. The axis Z_g points to the zenith along the normal of the local reference ellipsoid. Three axes form the right-hand rectangular coordinate system (E-N-U) (Fig. 2.7). The Chinese traditional conception of six directions (i.e., Up, Down, Fore, Behind, Left and Right—"六合 liu he") is the geographical coordinate system.

The three axes of the g system can be selected in different ways. The right-hand rectangular coordinate system can be constructed in the order of "north, west, up" (N-W-U) (Fig. 2.8) or "north, east and down" (N-E-D).

When the vehicle moves on the surface of the earth, the position of the vehicle relative to the earth changes constantly, and the g system of different places has different angular positions relative to the earth. That is, the motion of the vehicle will cause the local g system to rotate relative to the e system.

(2) Horizontal coordinate system (h system)

The so-called horizon is a plane with an observer on the Earth's surface at its center, which is the dividing line between the visible and invisible parts of the celestial

Fig. 2.8 Geographic
coordinate system (N-W-U)

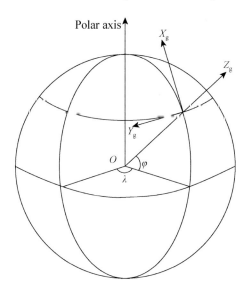

Fig. 2.8 Geographic coordinate system (N-W-U)

sphere. The celestial perpendicular line is consistent with the local gravity field through the observer. The two points above and below the observer's head are the zenith Z_h and the bottom Z_h'. Z_h is a celestial vertical axis, and the horizon plane is the base plane, which forms the horizontal coordinate system (Fig. 2.5).

(3) Tangent plane coordinate system (t system)

The t system is the fixed earth coordinate system. The tangent plane is the plane tangent to the geodetic reference ellipsoid, and the tangent point is the origin. This point is usually chosen as the landing site, navigation radar station, or some other convenient reference points. The axis X_t points east, the axis Y_t points north, and the axis Z_t points zenith and perpendicular to the reference ellipsoid, which form the right-hand rectangular coordinate system.

For stationary vehicles, the g system and the t system are identical. For the moving vehicle, the origin of the tangent plane is fixed, and the origin of the g system is the projection of the origin of the vehicle on the geoid. The t system is often used in local navigation, e.g., aircraft relative path navigation.

(4) Wandering azimuth coordinate system (w system)

The w system is defined on the basis of the geographical coordinate system. The origin of the w system is at the center of the vehicle; the axis X_w is perpendicular to the axis Y_w and the axis Z_w. The angle α, known as the drift azimuth, is between the axis Y_w in the local horizontal plane and the northern meridian circle (counterclockwise positive). The axis Z_w points to the zenith along the normal direction outside the ellipsoid α satisfy.

$$\alpha = -\lambda \sin \varphi \tag{2.2}$$

Fig. 2.9 Wandering azimuth
coordinate system

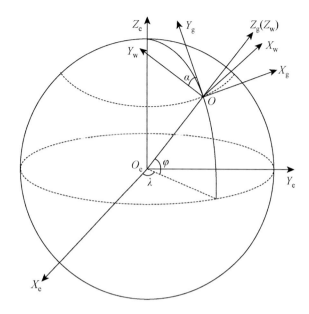

In the formula, λ and φ are the longitude and latitude of the vehicle center, respectively.

The use of the w system avoids the singularity of the g system in the polar region, so it is widely used in the polar navigation algorithm (Fig. 2.9).

(5) Greenwich grid coordinate system

The origin of the G system is the location of the vehicle. The plane parallel to the Greenwich meridian at the point is taken as the grid plane, and the horizontal plane where the vehicle is located is taken as the tangent plane. The intersection line between the grid plane and the tangent plane is defined as the grid north, and the angle between the grid north and the true north is represented as the grid azimuth σ. The zenith direction of the grid coincides with the geographic zenith direction, and the grid east is in the tangent plane and perpendicular to the grid north to form a right-hand rectangular coordinate system (Fig. 2.10).

Body Coordinate System (b System)

The b system is a general term for the aircraft coordinate system (aircraft), ship coordinate system (ship), star coordinate system (satellite), etc. (Fig. 2.11). Taking the body coordinate system of a ship as an example, the body coordinate system is fixed on the ship, and its origin is the mass center of the ship. Axis X_b points to starboard along the transverse axis of the ship, axis Y_b points to the bow along the longitudinal axis of the ship, and axis Z_b is perpendicular to the deck plane and points upward.

Fig. 2.10 Grid coordinate
system

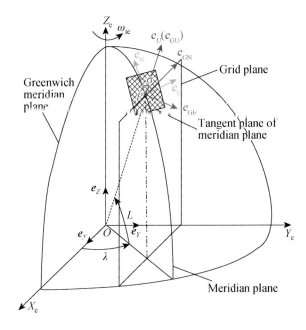

Fig. 2.11 Body coordinate
system (R-F-U)

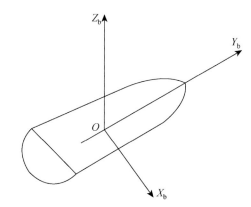

Other Coordinate Systems

In addition to the above coordinate systems, some other important coordinate systems
are used in the navigation list below.

(1) Navigation coordinate system (n system)

   The n system is the coordinate system selected by different navigation systems
   according to the principles when solving the navigation parameters.

(2) Platform coordinate system (p system)

For the inertial navigation system based on the stabilized platform, the p system is fixed with the physical platform, whose origin is located at the mass center of the platform, and the axes are consistent with the three axes indicated by the platform. For a strap down inertial navigation system, the p system is realized by the direction cosine matrix stored in the computer, so it is also called a "mathematical platform".

(3) Computing coordinate system (*c* system)

The c system refers to the navigation coordinate system determined by the computer navigation output. It has some error with the actual n system. The c system is often used in the analysis of navigation error.

(4) Gyroscope coordinate system (*gr* system)

The *gr* system is a coordinate system used to indicate the spatial orientation of the gyroscope rotor spin axis, which is fixed on the gyroscope internal frame. The origin is the center of the gyroscope bracket. The X_{gr} axis coincides with the rotor shaft but does not participate in the rotation of the rotor. The Y_{gr} axis coincides with the inner ring axis, and the Z_{gr} axis is always perpendicular to the X_{gr} and Y_{gr} planes. Three axes form the right-hand rectangular coordinate system.

2.2 The Coordinate System Transformation

In navigation research, an object is often abstracted as a coordinate system, and the motion of an object relative to another object is equivalent to the linear and angular motion of a coordinate system relative to another one. In the navigation system, one set of coordinate systems is connected to the studied object, and another set is connected to the selected reference space, the latter constituting the reference coordinate system of the former.

2.2.1 Coordinate Representation of the Vector

To define most motion parameters, including position, velocity, acceleration and angular velocity, three coordinate systems are involved:

(1) Body coordinate system (object frame), indicating the carrier, *b*;
(2) Reference coordinate system (reference frame), indicating the reference coordinate system of the movement, β;
(3) Projection coordinate system (resolving frame), indicating the measurement coordinate system, γ.

The *b* system cannot be the same as the β system; otherwise, there will be no relative motion. The γ system may be the *b* system, β system, or other coordinate systems. Just define the axis direction of the γ system and do not need to define its origin. The selection of the γ system does not affect the amplitude of the vector.

To fully describe the motion parameters, the above three coordinate systems should be clearly defined. The following symbols are indicated by:

$$x_{\beta\upsilon}^{\gamma} \tag{2.3}$$

Vector x can represent the Cartesian position, velocity, acceleration, angular velocity, etc., which describes the projected representation of the motion vector in frame b relative to frame β.

There are two main differences between the two rectangular coordinate systems. (1) The origins are different, that is, the origin of one coordinate system is displaced from another; (2) one coordinate system rotates relative to another. For example, the relationship between the g system ($OX_gY_gZ_g$) at a point on the Earth's surface and the E system ($OX_eY_eZ_e$).

Navigation involves many coordination transformation calculations, which include the transformation of points between different coordinate systems, such as transformations between rectangular coordinate systems and spherical coordinate systems, and various angle transformations between rectangular coordinate systems.

2.2.2 Position Transformation in the Coordinate System

The navigation system often transforms the position parameters between the spherical coordinate system and the rectangular coordinate system, which is as follows:

(1) Transform the relationship between spherical coordinates and rectangular coordinates (Fig. 2.12):

Fig. 2.12 Schematic diagram of the spherical coordinates

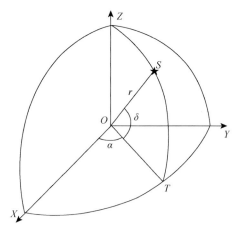

$$\begin{bmatrix} x \\ y \\ z \end{bmatrix} = r \begin{bmatrix} \cos \delta \cos \alpha \\ \cos \delta \sin \alpha \\ \sin \delta \end{bmatrix} \tag{2.4}$$

(2) Transform the relationship between rectangular coordinates and spherical coordinates:

$$
\begin{aligned}
r &= \sqrt{x^2 + y^2 + z^2} \\
\alpha &= \arctan \frac{y}{x} \\
\delta &= \arctan \frac{z}{\sqrt{x^2 + y^2}}
\end{aligned}
\tag{2.5}
$$

Position transformation can be commonly used in the e system. Let us suppose a point S on the Earth's surface in the e system, obtain $r_S = (x_S, y_S, z_S)$; the latitude and longitude coordinates are $P(\lambda, L)$; the distance from the point to the Earth's center is called the center radius, namely, $r_S = |r_S|$, and the transformation formula between the Earth's rectangular coordinates and spherical coordinates is as follows.

Transform the relationship between spherical coordinates to rectangular coordinates:

$$
\begin{cases}
x_s = R_N \cos L \cos \lambda \\
y_s = R_N \cos L \sin \lambda \\
z_s = R_N (1 - e)^2 \sin L
\end{cases}
\tag{2.6}
$$

where R_N is the radius of the prime vertical circle and e is the eccentricity of the reference ellipsoid.

Transform the relationship between rectangular coordinates to spherical coordinates:

$$
\begin{cases}
\lambda = arctan\dfrac{y}{x} \\
L = \arctan \left[\dfrac{1}{(1 - e^2)} \dfrac{z}{\sqrt{x^2 + y^2}} \right]
\end{cases}
\tag{2.7}
$$

2.2.3 Angle Relationship Representation Between Coordinate Systems

Determining the angular position and conversion relationship between two different coordinate systems is the fundamental problem of navigation. Navigation uses the g, b, p and c coordinate system, etc. In many cases, the origins of these frames are often approximately treated as the same, with no relative displacement, while mainly studying the relative angular variation. In the coordinate system transformation with different origins, the relative displacement between the origins can also be reflected by the angular position relationship between the coordinate systems. For example, the angular relationship between the e coordinate system and the g system can be determined by the angles λ and φ, which determine the position of the origin of the g coordinate system in the e coordinate system. Therefore, the angular relationship between the two coordinate systems is the key to analyzing the coordinate angular interconversion. The main methods to describe the angular position of the coordinate system are the directional cosine, Euler angle, quaternion and rotating vector methods.

Directional Cosine Matrix

The directional cosine method is a method of representing the coordinate transformation matrix by the directional cosine of the vector. The coordinate transformation matrix from coordinate system a to coordinate system b can be represented by a 3×3 azimuthal cosine matrix, i.e.,

$$C_a^b = \begin{bmatrix} C_{11} & C_{12} & C_{13} \\ C_{21} & C_{22} & C_{23} \\ C_{31} & C_{32} & C_{33} \end{bmatrix} \tag{2.8}$$

The various elements of the matrix are shown below. The unit vector of axes x_a, y_a, z_a can be represented by i_a, j_a and k_a. The unit vector of axes x_b, y_b, z_b can be represented by i_b, j_b, and k_b. Let any vector r be represented in i_a, j_a, k_a and i_b, j_b, k_b, respectively:

$$\begin{aligned} r^a &= x_a i_a + y_a j_a + z_a k_a \\ r^b &= x_b i_b + y_b j_b + z_b k_b \end{aligned} \tag{2.9}$$

The projection of vector r on the three axial vectors of coordinate system b is as follows:

$$\begin{aligned} x_b &= r^a \cdot i_b = (i_a \cdot i_b)x_a + (j_a \cdot i_b)y_a + (k_a \cdot i_b)z_a \\ y_b &= r^a \cdot j_b = (i_a \cdot j_b)x_a + (j_a \cdot j_b)y_a + (k_a \cdot j_b)z_a \\ z_b &= r^a \cdot k_b = (i_a \cdot k_b)x_a + (j_a \cdot k_b)y_a + (k_a \cdot k_b)z_a \end{aligned} \tag{2.10}$$

In Eq. (2.10), the transformation coefficients of the dot-product form $(i_a \, i_b)$ are the direction cosine, which can be obtained from the following form:

$$i_a \cdot i_b = |i_a||i_b|cos\theta_{i_b i_a} = \cos\theta_{i_b i_a} \tag{2.11}$$

where $\cos\theta_{i_b i_a}$ is the cosine between the two unit vectors i_a and i_b.

The above transformation equations can be further written in matrix form as follows:

$$\begin{bmatrix} x_b \\ y_b \\ z_b \end{bmatrix} = \begin{bmatrix} i_a \cdot i_b & j_a \cdot i_b & k_a \cdot i_b \\ i_a \cdot j_b & j_a \cdot j_b & k_a \cdot j_b \\ i_a \cdot k_b & j_a \cdot k_b & k_a \cdot k_b \end{bmatrix} \begin{bmatrix} x_a \\ y_a \\ z_a \end{bmatrix} \tag{2.12}$$

let

$$C_a^b = \begin{bmatrix} i_b \cdot i_a & i_b \cdot j_a & i_b \cdot k_a \\ j_b \cdot i_a & j_b \cdot j_a & j_b \cdot k_a \\ k_b \cdot i_a & k_b \cdot j_a & k_b \cdot k_a \end{bmatrix} \tag{2.13}$$

Namely, formula (2.12) can be written as

$$r^b = C_a^b r^a \tag{2.14}$$

Similarly, we can obtain

$$r^a = C_b^a r^b \tag{2.15}$$

$$C_b^a = \begin{bmatrix} i_b \cdot i_a & j_b \cdot i_a & k_b \cdot i_a \\ i_b \cdot j_a & j_b \cdot j_a & k_b \cdot j_a \\ i_b \cdot k_a & j_b \cdot k_a & k_b \cdot k_a \end{bmatrix} \tag{2.16}$$

Nine elements in C_a^b and C_b^a are all directional cosines between the axes of two coordinate systems, which reflects the angular relationship between two coordinate systems, called the directional cosine matrix from coordinate system a to b, C_a^b, and called the directional cosine matrix from coordinate system b to a, C_b^a.

Available by formulas (2.13) and (2.16):

$$(C_a^b)^{-1} = C_b^a = (C_a^b)^T \tag{2.17}$$

The directional cosine matrix is an orthogonal matrix. It is transitive, which enables the transformation between multiple coordinate systems easily. For example, in the aforementioned problem, if there is a third coordinate system $x_c y_c z_c$ and the directional cosine matrix of the coordinate system $x_b y_b z_b$ to the coordinate system $x_c y_c z_c$ is C_b^c, then the directional cosine matrix of the coordinate system $x_a y_a z_a$ to the coordinate system $x_c y_c z_c$ can be expressed as

$$C_a^c = C_b^c C_a^b \tag{2.18}$$

Fig. 2.13 Angular relation
of the coordinate system

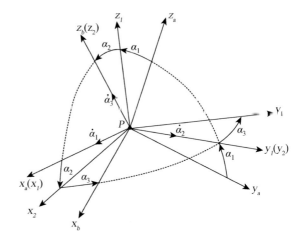

Euler Angle

The directional cosine matrix between two three-dimensional rectangular frames
has nine elements, depending on the properties of the orthogonal matrix, and the
quadratic sum of the three elements of the orthogonal matrix is one. There are actually
six constraints, and only three elements are independent. This shows that the angular
relationship between any two three-dimensional rectangular coordinate systems can
be completely described by three independent rotation angles, which are called Euler
angles.

By referring to the rotation of the coordinate system, we can define the transformation matrix between the orthogonal coordinate systems based on the rotation
angle (i.e., the Euler angle). In other words, the first reference frame $x_a y_a z_a$ rotates
3 times to produce the second reference frame $x_b y_b z_b$: first rotate around the x_a-axis
with the angle α_1; then rotate around the new y-axis with the angle α_2; finally rotate
around the new z-axis with the angle α_3, and obtain the second reference frame
$x_b y_b z_b$ (Fig. 2.13).

Quaternion

(1) Quaternion Definition

A quaternion is a number composed of four elements:

$$Q(q_0, q_1, q_2, q_3) = q_0 + q_1 i + q_2 j + q_3 k \qquad (2.19)$$

where q_0, q_1, q_2 and q_3 are real numbers; i, j, k are mutual orthogonal unit vectors
and the virtual units of $\sqrt{-1}$, which are embodied in the following quaternionic
multiplication relationship:

$$
\left.
\begin{array}{lll}
i \otimes i = -1, & j \otimes j = -1, & k \otimes k = -1 \\
i \otimes j = k, & j \otimes k = i, & k \otimes i = j \\
j \otimes i = -k, & k \otimes j = -i, & i \otimes k = -j
\end{array}
\right\}
\qquad (2.20)
$$

where $\otimes$ indicates the quaternionic multiplication.

Quaternions, proposed by the Irish mathematician Hamilton, can be regarded as a super complex number consisting of the real part scalar q_0 and the virtual part three-dimensional vectors. The three axes i, j, and k are orthogonal to the virtual part and can actually be regarded as a vector in a four-dimensional space. Similar to the mathematical significance of complex number multiplication, quaternionic multiplication can easily solve the complex vector rotation transformation problem in three-dimensional space.

(2) The Relationship Between the Quaternions and the Directional Cosine Matrix

The reference coordinate system β is provided, and the unit vectors of the x_0, y_0, and z_0 axes are i_0, j_0, and k_0, respectively. The axes $x_b y_b z_b$ of the body coordinate system b rotate around the fixed point O. To explain the relative angular position and rotation relationship of the body, point A on the body and turning point O are taken to obtain the position vector OA (Fig. 2.14). The position vector can simplify the angular position of the body.

Fig. 2.14 Equivalent rotation of the rigid body

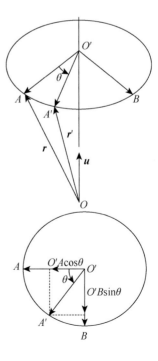

Let the b system rotate relative to the β system; the initial position vector $OA = r$. After time t, the position vector $OA' = r'$. According to Euler's theorem, and only considering the angular position of the b system at time 0 and t, the rotation of the b system from position A to position A' can be achieved directly by rotating around the axis u (unit vector) with the angle θ, where $u = l i_0 + m j_0 + n k_0$. Position vectors make a conical motion, with A and A' on the same circle and r and r' on the same conical plane.

Let

$$
\begin{cases}
q_0 = \cos\dfrac{\theta}{2} \\[2mm]
q_1 = l \sin\dfrac{\theta}{2} \\[2mm]
q_2 = m \sin\dfrac{\theta}{2} \\[2mm]
q_3 = n \sin\dfrac{\theta}{2}
\end{cases}
\tag{2.21}
$$

In addition, construct quaternions based on q_0, q_1, q_2, and q_3:

$$
\begin{aligned}
Q &= q_0 + q_1 i_0 + q_2 j_0 + q_3 k_0 = \cos\frac{\theta}{2} + (l i_0 + m j_0 + n k_0)\sin\frac{\theta}{2} \\
&= \cos\frac{\theta}{2} + u^\beta \sin\frac{\theta}{2}
\end{aligned}
\tag{2.22}
$$

where u^β is the rotation axis direction and θ is the turning angle. The following conclusions can be obtained:

(a) Quaternion Q describes the fixed rotation of the body, that is, when only concerned about the angular position of the b system relative to β system, the b system can be considered to be formed by a disposable equivalent rotation of the system β directly; Q contains all the information of this equivalent rotation, and it can prove that the vectors r and r' through rotation satisfy the following relationship:

$$
r' = Q \cdot r \cdot Q^*
\tag{2.23}
$$

where $Q^* = q_0 - q_1 i_0 - q_2 j_0 - q_3 k_0$ and is the conjugate of the quaternions Q.

(b) It can be deduced that there is the following correspondence between quaternion Q and the coordinate transformation matrix from system b to system β.

$$C_b^\beta = \begin{bmatrix} 1 - 2(q_2^2 + q_3^2) & 2(q_1q_2 - q_0q_3) & 2(q_1q_3 + q_0q_2) \\ 2(q_1q_2 + q_0q_3) & 1 - 2(q_1^2 + q_3^2) & 2(q_2q_3 - q_0q_1) \\ 2(q_1q_3 - q_0q_2) & 2(q_2q_3 + q_0q_1) & 1 - 2(q_1^2 + q_2^2) \end{bmatrix} \tag{2.24}$$

2.2.4 Common Coordinate System Transformation

The following lists the common coordinate transformations in navigation.

The Inertial Coordinate System—The Earth Coordinate System

From the ECI system $X_iY_iZ_i$ to the e system $X_eY_eZ_e$, only one rotation is needed:

$$C_i^e = \begin{bmatrix} \cos(\omega_{ie}t - \lambda_0) & \sin(\omega_{ie}t - \lambda_0) & 0 \\ -\sin(\omega_{ie}t - \lambda_0) & \cos(\omega_{ie}t - \lambda_0) & 0 \\ 0 & 0 & 1 \end{bmatrix} \tag{2.25}$$

Conversely,

$$C_i^e = \begin{bmatrix} \cos(\omega_{ie}t - \lambda_0) & -\sin(\omega_{ie}t - \lambda_0) & 0 \\ \sin(\omega_{ie}t - \lambda_0) & \cos(\omega_{ie}t - \lambda_0) & 0 \\ 0 & 0 & 1 \end{bmatrix} \tag{2.26}$$

where ω_{ie} is the earth rotation angular velocity, t is the navigation time, and λ_0 is the angle between the X axes of the ECI system and the Earth coordinate system.

Inertial Coordinate System – Geographic Coordinate System (N-E-D)

The inertial coordinate system $X_iY_iZ_i$ must pass through two consecutive rotations from the geographic coordinate system $X_gY_gZ_g$.
First,

$$[\lambda']z_i = \begin{bmatrix} \cos\lambda' & \sin\lambda' & 0 \\ -\sin\lambda' & \cos\lambda' & 0 \\ 0 & 0 & 1 \end{bmatrix} \tag{2.27}$$

Second,

$$\left[-\left(\frac{\pi}{2}+L\right)\right]_{Y_i'} = \begin{bmatrix} \cos\left(\frac{\pi}{2}+L\right) & 0 & \sin\left(\frac{\pi}{2}+L\right) \\ 0 & 1 & 0 \\ -\sin\left(\frac{\pi}{2}+L\right) & 0 & \cos\left(\frac{\pi}{2}+L\right) \end{bmatrix} = \begin{bmatrix} -\sin L & 0 & -\cos L \\ 0 & 1 & 0 \\ \cos L & 0 & -\sin L \end{bmatrix}$$

(2.28)

Thus, the coordinate transformation matrix from the i system to the g system is

$$
\begin{aligned}
C_i^g &= \left[-\left(\frac{\pi}{2}+L\right)\right]_{Y_i'} [\lambda']_{Z_i} \\
&= \begin{bmatrix} -\sin L & 0 & -\cos L \\ 0 & 1 & 0 \\ \cos L & 0 & -\sin L \end{bmatrix} \begin{bmatrix} \cos\lambda' & \sin\lambda' & 0 \\ -\sin\lambda' & \cos\lambda' & 0 \\ 0 & 0 & 1 \end{bmatrix} \\
&= \begin{bmatrix} -\cos\lambda'\sin L & -\sin\lambda'\sin L & -\cos L \\ -\sin\lambda' & \cos\lambda' & 0 \\ \cos\lambda'\cos L & \sin\lambda'\cos L & -\sin L \end{bmatrix}
\end{aligned}
$$

(2.29)

where L is geographical latitude and $\lambda' = L - L_0 + \omega_{ie}t$ is geographical longitude. Otherwise, the transformation matrix from the g system to the ECI system is

$$C_g^i = \begin{bmatrix} -\cos\lambda'\sin L & -\sin\lambda' & \cos\lambda'\cos L \\ -\sin\lambda'\sin L & \cos\lambda' & \sin\lambda'\cos L \\ -\cos L & 0 & -\sin L \end{bmatrix}$$

(2.30)

Earth Coordinate System—Geographic Coordinate System (N-E-D)

From the ECI system to the g system, it must go through two consecutive plane rotations, and we can obtain the "north, east and down" geographic coordinate system.
First,

$$[\lambda]_{Z_c} = \begin{bmatrix} \cos\lambda & \sin\lambda & 0 \\ -\sin\lambda & \cos\lambda & 0 \\ 0 & 0 & 1 \end{bmatrix}$$

(2.31)

Second,

$$
\begin{aligned}
\left[-\left(\frac{\pi}{2}+L\right)\right]_{Y_c'} &= \begin{bmatrix} \cos\left(\frac{\pi}{2}+L\right) & 0 & \sin\left(\frac{\pi}{2}+L\right) \\ 0 & 1 & 0 \\ -\sin\left(\frac{\pi}{2}+L\right) & 0 & \cos\left(\frac{\pi}{2}+L\right) \end{bmatrix} \\
&= \begin{bmatrix} -\sin L & 0 & -\cos L \\ 0 & 1 & 0 \\ \cos L & 0 & -\sin L \end{bmatrix}
\end{aligned}
$$

(2.32)

Thus, the coordinate transformation matrix from the ECI system to the g system is

$$
C_e^g = \begin{bmatrix} -\sin L & 0 & -\cos L \\ 0 & 1 & 0 \\ \cos L & 0 & -\sin L \end{bmatrix} \begin{bmatrix} \cos \lambda & \sin \lambda & 0 \\ -\sin \lambda & \cos \lambda & 0 \\ 0 & 0 & 1 \end{bmatrix}
$$
$$
= \begin{bmatrix} -\cos \lambda \sin L & -\sin \lambda \sin L & -\cos L \\ -\sin \lambda & \cos \lambda & 0 \\ \cos \lambda \cos L & \sin \lambda \cos L & -\sin L \end{bmatrix}
$$
(2.33)

where $\lambda = L - L_0 = \lambda' - \omega_{ie}t$ is the difference between the local and the origin geographical longitude (when $L_0 = 0$, λ is the geographical longitude).

Otherwise, the coordinate transformation matrix from the g system to the ECI system is

$$
C_g^e = \begin{bmatrix} -\cos \lambda \sin L & -\sin \lambda & \cos \lambda \cos L \\ -\sin \lambda \sin L & \cos \lambda & \sin \lambda \cos L \\ -\cos L & 0 & -\sin L \end{bmatrix}
$$
(2.34)

The Geographic Coordinate System—The Body Coordinate System

The attitude angles are the three angles between the body coordinate system and the geographic coordinate system. Their definitions are defined as follows.

(1) **Direction angle** (ψ): the angle between the longitudinal axis of body X_b and the north axis (N) in the horizontal plane, and clockwise is positive;
(2) **pitch angle** (θ): the angle between the longitudinal axis X_b of the body and the horizontal plane in the vertical plane, and elevation is positive;
(3) **Rolling Angle** (ϕ): the angle between the longitudinal axis of the body Y_b and the horizontal plane in the cross section, and the left lift is positive.

Note: According to the attitude angle definition, a geometric diagram between the b system $X_b Y_b Z_b$ and the g system (N-E-D) is drawn (Fig. 2.15). The transformation process from the g system (N-E-D) to the b system (F-R-D, "front, right and down") is introduced. According to the geometry relation of the Euler angle, the g system can obtain the b system by successively rotating with heading angle (ψ), pitch angle (θ) and roll angle (ϕ).

The first rotation:

$$
[\psi]_D = \begin{bmatrix} \cos \psi & \sin \psi & 0 \\ -\sin \psi & \cos \psi & 0 \\ 0 & 0 & 1 \end{bmatrix}
$$
(2.35)

Fig. 2.15 Definition of attitude

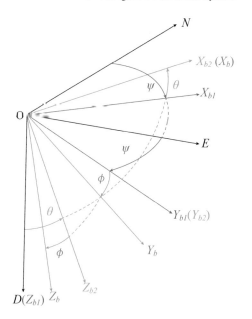

The second rotation:

$$[\theta]_{Yb1} = \begin{bmatrix} \cos\theta & 0 & -sin\theta \\ 0 & 1 & 0 \\ sin\theta & 0 & \cos\theta \end{bmatrix} \tag{2.36}$$

The third rotation:

$$[\phi]_{Xb2} = \begin{bmatrix} 1 & 0 & 0 \\ 0 & \cos\phi & sin\phi \\ 0 & -\sin\phi & \cos\phi \end{bmatrix} \tag{2.37}$$

Therefore, the coordinate transformation matrix from the g system to the b system is

$$C_g^b = [\phi]_{Xb2}[\theta]_{Yb1}[\psi]_D$$

$$= \begin{bmatrix} 1 & 0 & 0 \\ 0 & \cos\phi & \sin\phi \\ 0 & -\sin\phi & \cos\phi \end{bmatrix} \begin{bmatrix} \cos\theta & 0 & -\sin\theta \\ 0 & 1 & 0 \\ \sin\theta & 0 & \cos\theta \end{bmatrix} \begin{bmatrix} \cos\psi & \sin\psi & 0 \\ -\sin\psi & \cos\psi & 0 \\ 0 & 0 & 1 \end{bmatrix}$$

$$= \begin{bmatrix} \cos\psi\cos\theta & \sin\psi\cos\theta & -\sin\theta \\ -\sin\psi\cos\phi + \cos\psi\sin\theta\sin\phi & \cos\psi\cos\phi + \sin\psi\sin\theta\sin\phi & cos\theta\sin\phi \\ \sin\psi\sin\phi + \cos\psi\sin\theta\cos\phi & -\cos\psi\sin\phi + \sin\psi\sin\theta\cos\phi & cos\theta\cos\phi \end{bmatrix}$$

$$\tag{2.38}$$

Otherwise, the coordinate transformation matrix from the b system to the g system is

$$C_b^g = \begin{bmatrix} \cos\psi\cos\theta & -\sin\psi\cos\phi + \cos\psi\sin\theta\sin\phi & \sin\psi\sin\phi + \cos\psi\sin\theta\cos\phi \\ \sin\psi\cos\theta & \cos\psi\cos\phi + \sin\psi\sin\theta\sin\phi & -\cos\psi\sin\phi + \sin\psi\sin\theta\cos\phi \\ -\sin\theta & \cos\theta\sin\phi & \cos\theta\cos\phi \end{bmatrix}$$

(2.39)

The Euler angle can be calculated based on the elements in the directional cosine matrix C_g^b:

$$\tan\phi = \frac{C_{23}}{C_{33}} = \frac{\sin\phi\cos\theta}{\cos\phi\cos\theta} = \frac{\sin\phi}{\cos\phi}, \quad \phi = \tan^{-1}\left(\frac{C_{23}}{C_{33}}\right)$$

(2.40)

$$\tan\psi = \frac{C_{12}}{C_{11}} = \frac{\sin\psi\cos\theta}{\cos\psi\cos\theta} = \frac{\sin\psi}{\cos\psi}, \quad \psi = \tan^{-1}\left(\frac{C_{12}}{C_{11}}\right)$$

(2.41)

$$-\tan\theta = \frac{C_{13}}{\sqrt{1 - C_{13}^2}} = \frac{-\sin\theta}{\sqrt{1 - \sin^2\theta}} = \frac{-\sin\theta}{\cos\theta}, \quad \theta = \tan^{-1}\left(\frac{-C_{13}}{\sqrt{1 - C_{13}^2}}\right)$$

(2.42)

The attitude information of the body can be calculated from the coordinate transformation matrix C_b^g.

The transformation relationship between the various coordinate systems is shown in Fig. 2.16.

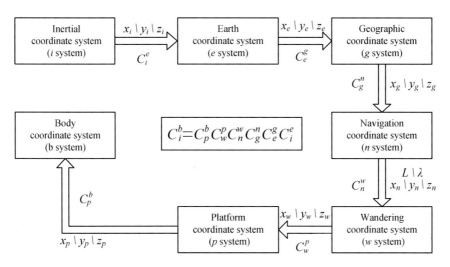

Fig. 2.16 Transform relationships between the coordinate systems in the navigation system

2.3   Geodetic Coordinate System

2.3.1   *Mathematical Description of the Earth's Shape*

At present, humans mainly focus on navigation near the surface of the earth. How can the Earth be described as accurately as possible in the coordinate system?

The earth is an approximate sphere with a complex shape, not a standard sphere. There are mountains, deep valleys and plains on land and reefs and trenches in the ocean. It is a very complex and irregular surface. Therefore, the natural surface of the earth is not a mathematical surface and cannot be directly calculated [3].

For the convenience of scientific research, the irregular natural form of the earth can be replaced by a form that can be expressed mathematically. Assuming that the sea water in the ocean is in a completely static and balanced state, i.e., Without the influence of ocean currents, tides, wind and waves, the calm sea surface is called a geoid. The geoid is a level that corresponds to the hypothetical and fully equilibrium sea level and is perpendicular to the plumb lines everywhere. If it extends to the mainland to form a continuous level, it will extend to the entire surface of the earth, which is called the geoid ellipsoid (Figs. 2.17 and 2.18).

Fig. 2.17 Geoid diagram

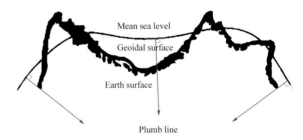

Fig. 2.18 Geoid ellipsoid

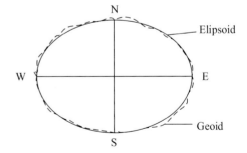

Earth Sphere

In general, applications, the shape of the earth is regarded as a sphere with radius R, that is, the first approximation of the earth, while the ellipsoid is regarded as its second approximation. The average radius of the Earth's sphere is $R \approx 6371.02 \pm 0.05$ km, and the earth rotation angular velocity is $\omega \approx 7.29 \times 10^{-5}$ rad/s.

Earth Ellipsoid

In more accurate navigational calculations, it is necessary to assume the earth as a slightly smooth ellipsoid (Fig. 2.19). The ellipsoid is composed of ellipses $P_N Q P_s Q'$ rotating around its minor semi axis $P_N P_s$.

The important parameters of the Earth's ellipsoid are the major semi axis a, minor semi axis b, oblate c and eccentricity e. The interrelations between them are:

$$c = \frac{a-b}{a}; e = \frac{\sqrt{a^2 - b^2}}{a}; \tag{2.43}$$

In different historical periods, the measured results are different, so the parameters of the ellipsoid solid are different. Table 2.1 lists the main parameters of various ellipsoids once commonly used in the world.

Fig. 2.19 A schematic diagram of the earth's ellipsoid

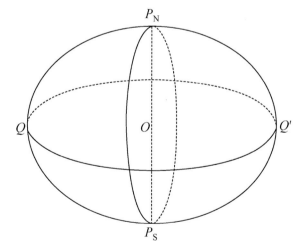

Table 2.1 Earth ellipsoid parameters

Ellipsoid	Year	Major semi axis (m)	Oblatec	Main user countries
Bassel	1841	6,377,397.155	1/299.1528	Germany, Switzerland, Japan
Clarke	1866	6,378,206,4	1/294.978	USA, Canada, Mexico
Hayford	1910	6,378,388	1/297.0	USA, West European
Krasovsky	1940	6,378,245	1/298.3	USSR, East Europe, China
WGS-84	1984	6,378,137	1/298.257 22	USA

2.3.2 Common Geodetic Coordinate System

The geodetic coordinate system uses longitude, latitude and height to describe spatial position (B, L, H). The definition of a geodetic coordinate system includes the origin of the coordinate system, the direction and scale of the three coordinate axes, and four basic parameters of the Earth's ellipsoid (the major semi axis a, the second-order spherical harmonic coefficient of the Earth's gravity field J_2, the gravitational constant GM and the angular velocity of the Earth's rotation ω). Due to historical and technological reasons, China has established and used many different geodetic coordinate systems in different periods, and geodetic coordinate systems have also changed from the reference ellipsoid centric coordinate system to the geocentric coordinate system. The geodetic coordinate system commonly used in China is briefly introduced here.

1980 Xi'an Coordinate System

In 1980, a reference-ellipsoid-centric coordinate system was established through the adjustment of the National celestial geodetic network, also known as the Xi'an coordinate system. The origin of the Xi'an coordinate system was located in Yongle Town, Jinghe County, Shaanxi Province.

The four geometric and physical parameters of the ellipsoid parameters adopted in the coordinate system were recommended by the IAG in 1975. The minor semi axis of the ellipsoid is parallel to the rotation axis of the earth, and the starting meridian plane is parallel to the Greenwich mean astronomical meridian plane. The geoid of the ellipsoid is in good agreement in China.

The coordinate system uses four basic ellipsoidal parameters, including both geometric and physical parameters. The values are recommended by the 16th Congress of IUGG 1975.

Major semi axis	$a = 6378140\,\mathrm{m}$
Gravitational constant	$GM = 3.986005 \times 10^{14}\,\mathrm{m}^3/\mathrm{s}^2$
Dynamic shape factor	$J_2 = 1.08263 \times 10^{-3}$
Earth rotation angular velocity	$\omega = 7.292115 \times 10^{-5}\,\mathrm{rad}/\mathrm{s}$

The coefficient of the second-order main sphere function of the J_2 earth gravity field is a function of oblateness. The elevation system is based on the average sea level of China's Yellow Sea in 1956. Other coordinate system constants can be derived from the above four basic constants.

Compared with the Beijing coordinate system in 1954, the Xi'an coordinate system in 1980 has the following advantages: (1) the origin of the coordinates is located in China; and (2) the reference ellipsoid is more suitable. The minor semi axis of the ellipsoid points to the polar origin JYD1968.0. (3) The ellipsoid is in good agreement with China's geoid, with an average difference of 10m across the country. (4)The accuracy of geocentric coordinates obtained from the Xi'an coordinate system in 1980 has been improved through different types of mathematical models and their conversion parameters.

However, the following problems still exist in the Xi'an coordinate system in 1980: (1) It is still a two-dimensional coordinate system and cannot provide high-precision three-dimensional coordinates. (2) The ellipsoid used is approximately 3 m larger than that used in the satellite positioning system, which may cause errors in the surface length with the magnitude 5×10^{-7}. (3) The ellipsoidal minor semi axis points to the polar origin JYD1968.0, which is different from the general international terrestrial coordinate system such as ITRS or the direction of the ellipsoidal short axis such as WGS84 (BIH1984.0) used in GPS (Fig. 2.20).

Fig. 2.20 WGS84
coordinate system definition

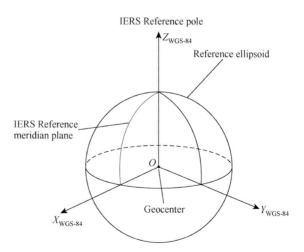

WGS84 International Coordinate System

The full name of the WGS-84 coordinate system is World Geodatic System-84. It is the coordinate system adopted by GPS. The ephemeris parameters of GPS are based on this coordinate system. It is the most widely used global geodetic reference system in the field of navigation and surveying.

The WGS-84 coordinate system is an ECEF system. The coordinate origin is located at the center of mass of the earth, the Z axis points to the agreed polar direction of the earth defined by BIH 1984.0, the X axis points to the intersection of the starting meridian plane and the equator of BIH 1984.0, and the Y axis forms a right-handed system with the X axis and the Z axis.

The four basic ellipsoidal parameters adopted by the rotating reference ellipsoid are as follows:

Major semi axis	$a = 6378137\,\text{m}$
Gravitational constant	$GM = 3.986004418 \times 10^{14}\,\text{m}^3/\text{s}^2$
Dynamic shape factor	$J_2 = 1.08263 \times 10^{-3}$
Earth rotation angular velocity	$\omega = 7.292115 \times 10^{-5}\,\text{rad/s}$

The WGS84 coordinate system was developed by NIMA (National Image and Mapping Agency) and its predecessor, the Defense Mapping Agency (DMA) of the United States Department of Defense, from the initial world geodetic coordinate system WGS60 and developed on the basis of the subsequent WGS66 and WGS72. Through the use of GPS, the implementation of the WGS84 reference frame has made great progress. The WGS84 coordinate system is realized by calculating GPS tracking stations with absolute accurate coordinates.

PZ-90 Coordinate System

The PZ-90 coordinate system was adopted by GLONASS in 1993. It belongs to the ECEF system. The coordinate reference frame of the GLONASS system is realized by a series of coordinates of tracking stations. Before 1993, GLONASS adopted the Soviet Geodetic System (SGS-85) in 1985. After 1991, the GLONASS system control center accepted the suggestion of the Russian Map Bureau, improved the longitude orientation and Z-axis origin position of SGS-85, and established the PZ-90 (Parametry Zelmy 1990; English translation: Parameters of the Earth) coordinate system. PZ-90 is based on Doppler observations, satellite laser ranging (SLR), GEO-IK altimetry satellite and GLONASS satellite radar ranging and other large amounts of data, which are calculated through the combined adjustment of the ground network and space network. Since 1993, GLONASS has changed to the PZ-90 coordinate system [4].

The coordinate system PZ-90 adopted by GLONASS is defined as follows:

(1) The origin of the coordinates is located at the mass center of the earth;

(2) The Z axis points to the conventional terrestrial pole recommended by the International Earth Rotation Service (IERS);
(3) The X-axis points to the zero meridian intersection defined by the Earth's equator and the International Bureau of Time (BIH);
(4) The Y-axis satisfies the right-hand coordinate system.

The reference ellipsoid parameters and other parameters used in the PZ-90 coordinate system are listed as follows.

Major semi axis	$a = 6378136$ m
Gravitational constant	GM=3.9860044×10^{14} m^3/s^2
Dynamic shape factor	$J_2 = 1.0825257 \times 10^{-3}$
Earth rotation angular velocity	$\omega = 7.292115 \times 10^{-5}$rad/s

Although PZ-90 is a geocentric geostationary coordinate system and its definition is the same as that of the International Terrestrial Reference Frame (ITRF), there are some differences between the defined coordinate system and the actual coordinate system due to the inevitable positioning and measurement errors of the tracking station. In fact, PZ-90 differs from WGS 84 and ITRF in origin, orientation and scale of coordinates. The difference in coordinates between PZ-90 and WGS-84 on the Earth's surface can reach 20 mm.

CGCS2000 China Geodetic Coordinate System

Since July 1, 2008, China has launched the National Geodetic Coordination System (CGCS 2000) and used 10 years to complete the conversion from the current geodetic coordinate system. The CGCS 2000 national geodetic coordinate system is adopted in the Beidou navigation satellite system in China.

The origin of the CGCS 2000 national geodetic coordinate system is the mass center of the Earth, including the ocean and atmosphere; the Z axis of the coordinate system points from the origin to the Earth reference pole of epoch 2000.0, and the direction of the epoch is calculated from the initial direction of epoch 1984.0 given by the International Bureau of Time. The directional time evolution ensures that no residual global rotation occurs relative to the crust, and the X axis points from the origin to the intersection between the reference meridian and the equatorial surface of the Earth (epoch 2000.0). The Y axis, Z axis and X axis constitute a right-hand orthogonal coordinate system, using the scale in the sense of GR.

The ellipsoid parameters adopted in CGCS2000 are as follows:

Major semi axis	$a = 6378137$ m
Gravitational constant	$GM = 3.986004418 \times 10^{14}$ m^3 /s^2
Dynamic shape factor	$J_2 = 1.082629832258 \times 10^{-3}$

(continued)

(continued)

Major semi axis	$a = 6378137$ m
Earth rotation angular velocity	$\omega = 7.292115 \times 10^{-5}$ rad/s
Oblate	$f = 1/298.257222101$

The CGCS2000 national geodetic coordinate system is basically consistent with the International Terrestrial Reference Coordination System (ITRF). ITRFis more accurate than WGS84. In addition, the GTRF is the geodetic coordinate system used by the GTRF, which also fully used for reference in the ITRF.

Comparison Between Geocentric Coordinate System and Reference-Ellipsoid-Centric Coordinate System

WGS-84, PZ-90, CGCS2000, GTRF and ITRF are all geocentric coordinate systems, which are quite different from the 1980 Xi'an coordinate system in China. The differences between them are mainly reflected in the following aspects:

(1) Different ellipsoidal positioning methods. The reference-ellipsoid-centric coordinate system is used to study the local spherical shape. Under the principle of minimizing the correction of the land survey data to the ellipsoid, the ellipsoid coordinate system, which is most consistent with the selected local geoid, is not conducive to the study of global shape and plate motion and is unable to establish a unified global geodetic coordinate system. WGS-84, PZ-90 and CGCS2000 are geocentric coordinate systems and the ellipsoid positioning is the closest to the global geoid.

(2) Implementation technology is different. The reference-ellipsoid-centric coordinate system adopts traditional geodetic means, i.e., measuring the distance and direction between landmarks. The position of each point relative to the starting point is obtained by the adjustment method, and the coordinates of each point under the paracentric system are determined. The WGS-84, PZ-90 and CGCS 2000 frameworks are used to obtain the geocentric coordinates of stations under the ITRF framework through space geodetic observation technology.

(3) Different origins. The origin of the reference-ellipsoid-centric coordinate system is quite different from the mass center of the earth, and the origin of the geocentric coordinate system is located in the mass center of the earth.

(4) Different accuracies. Due to the limitation of objective conditions at that time, the reference-ellipsoid-centric coordinate system lacks high-precision external control, and its long-distance accuracy is low. Today, the reference-ellipsoid-centric coordinate system has difficulty meeting the needs of users. The precision of CGCS 2000 is 10 times higher than that of the reference-ellipsoid-centric coordinate system, and the relative precision can reach $10^{-7} \sim 10^{-8}$.

In other words, the geocentric coordinate system is conducive to the maintenance and rapid updating of the coordinate system by modern space technology, the coordinate determination of high-precision geodetic control stations, and the improvement of mapping efficiency. It can better clarify various geographic and physical phenomena on the earth, especially the movement of space objects. Thus, the geocentric coordinate system has been the general trend.

Appendix 2.1: Questions

1. How do the rectangular coordinate system, spherical coordinate system and cylindrical coordinate system apply in navigation? What coordinate system is the latitude and longitude information and Mercator projection commonly used in navigation?
2. How can we understand the inertial coordinate system? Why are there many different definitions of inertial coordinate systems in navigation? Among the sun-center inertial coordinate system, earth-center inertial coordinate system, geographic coordinate system and body coordinate system, what are the coordinate systems not affected by the Earth's rotation?
3. Please list the definitions, English terms and abbreviations of the various coordinate systems introduced in the textbook.
4. What are the common coordinate transformation methods in navigation? What are the main mathematical methods by which to represent the angular relationship between coordinate systems in navigation?
5. What are the first and second approximations to the mathematical description of the earth shape? What is the geoid? How can the shape of the real earth surface be accurately understood?
6. What are the main elements of the definition of the geodetic coordinate system? The differences in the parameters in Xi'an 1980, WGS84, PZ 90, CGCS2000 and other geodetic coordinates are compared.
7. What are the main differences between geocentric coordinate systems and paracentric coordinate systems?
8. How can we understand the great significance of the concept of introducing the coordinate system in the development of modern science? Is there an ultimate inertial coordinate system?

Appendix 2.2: The Significance of Describing the World Based on the Coordinate System

Different coordinate systems show different human concerns, which to some extent reflects the relativity and generality of people's cognitive process. A large number of coordinate systems produce the abstraction of the objective world with a particular way of professional thinking. The correlation between different coordinate systems constitutes a dynamic and complex world outlined in the "description" of the coordinate frame. All kinds of complex carrier movement relationships are reflected in the changing and complex linear and angular movements between different coordinate systems.

A deep understanding of coordinate systems can expand the understanding of navigation and of the key foundation of modern physics. In fact, describing the world in the space–time frame of the coordinate systems is a major advance in the development of modern science. In particular, the introduction of the mathematical analysis method based on coordinate systems by Descartes has transformed the ancient eastern concepts, such as "liuhe ("六合", six directions of the world)" and "shifang ("十方", ten directions of the world)", into an abstract but accurate space–time framework, which changed the human understanding of the world and laid the foundation of modern scientific research paradigm. Knowing this can help readers to deepen their thinking and understanding of scientific thinking.

In addition, a profound and fundamental question of physics is whether there is an ultimate inertial coordinate system in the objective world. To answer this question involves the Newtonian system of physics and the Einstein system of physics. Seeking answers to this problem encourages readers to fully understand the inertial coordinate system, and more importantly, to expand their thinking dimension with the rational thinking in the space–time frame, and to further address the "ontological" problem.

Chapter 3
PNT Parameters

Positioning, navigation and timing (PNT) parameters are the basic spatial–temporal reference parameter information of ships, aircraft and other carriers. On July 31, 2020, China's Beidou-III system was officially started for service, and the state clearly put forward the new goal of building the Beidou comprehensive PNT system. In the future, a more ubiquitous, integrated and intelligent comprehensive PNT system will be built. In short, there are three core objectives of the PNT system, namely, positioning service, navigation service and timing service. The PNT system is a general term for all kinds of infrastructure, equipment, technology, operation and maintenance, management, application and policy required for accurate spatial and temporal information acquisition, carrier mobility guidance and information decision support within the spatial and spatial regions concerned. PNT-related technology has become a research hotspot around the world.

However, because the PNT system is new, some conceptual terms have not yet reached consensus and have no unified definition. With the development of technology, the connotation and extension of the concept of navigation information are also changing quietly, and users have also changed profoundly. However, conceptual consistency is the premise of professional research. This chapter provides a systematic introduction to the definition of and differences in PNT parameters that are closely related to navigation.

3.1 Positioning Parameters

Positioning is used to determine the spatial position of the vehicle. From the point of view of traditional navigation, positioning is a function of navigation; however, navigation is one of many applications of positioning. Other applications of positioning include surveying, mapping, tracking, surveillance, mechanical control, construction engineering, vehicle testing, geoscience, intelligent transportation systems and

© Science Press 2024
H. Bian et al., *Essentials of Navigation*, https://doi.org/10.1007/978-981-99-5636-4_3

location-based services. Therefore, there is a conceptual overlap between positioning and navigation, especially when describing PNT problems.

Positioning information is one of the most important information. Chapter II introduced that the position of a point on the Earth's surface can be represented by a rectangular coordinate system and spherical coordinate system, and there is a clear conversion formula between them. In near-Earth surface navigation systems such as ships and aircraft, the vehicle position is mainly expressed by spherical coordinates (longitude, latitude and altitude). This section mainly discusses the spherical coordinate parameters of positioning.

3.1.1   Latitude

In geographic coordinates, the intersection line between the plane perpendicular to the Earth's axis and the Earth's surface is called the latitude circle, or the small circle parallel to the equator (FGF'). An arc in a latitude circle is called a latitude line. The largest latitude circle is the equator (Fig. 3.1).

The latitude of a point on the earth is defined as the angle between the vertical line of the point and the equatorial plane. However, because the Earth can be approximated as a revolving ellipsoid, there are many definitions of vertical lines and corresponding latitudes.

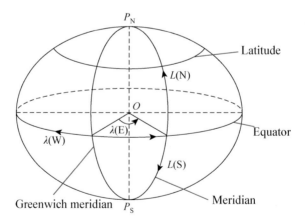

Fig. 3.1 Geographic coordinate sketch

Geocentric Latitude

The connection from a point O on the ellipsoid of the earth to the center of the earth is called the vertical line of the earth center (OO_e). The angle between OO_e and the equatorial plane is called the geocentric latitude of point O. The north latitude is calculated from the equator to the north, marked by "N", and the south latitude is marked by "S". When calculating, the north latitude is positive, and the south latitude is negative, e.g., $L = 41° 54.2'$ N.

$$\sin L'_S = \frac{z_S}{R_S} \tag{3.1}$$

Celestial Latitude

The gravity vector of a point on the ellipsoid of the earth is called the celestial vertical line. The angle between the gravity vector of a point on the ellipsoid of the earth and the equatorial plane is called the celestial latitude. The traditional latitude measurement of a certain point mainly relies on the determination of the local vertical line by the direction of the lead hammer, so the latitude measured is the celestial latitude.

Celestial latitude measurement has two shortcomings: (1) because of the existence of local gravity anomalies, there are many points on the same meridian with the same celestial latitude; and (2) the movement of the Earth's poles will lead to minor changes in latitude at any point on the earth.

Geographical Latitude

The normal line of a point on the elliptic surface of the earth is called the geographic vertical line. (Fig. 3.2) The angle between the normal line of a point on the elliptic surface of the earth and the equatorial plane is called geographic latitude. Geographic latitude is an important parameter in the geodetic domain, so it is also called geodetic latitude. The latitude used in the inertial navigation system is geographic latitude.

Since the geographic latitude is defined by the surface normal, it can be obtained by calculating the surface gradient of the ellipsoid [5].

$$\tan L_S = \frac{z_S}{(1 - e^2) r_S} = \frac{z_S}{(1 - e^2)\sqrt{x_S^2 + y_S^2}} \tag{3.2}$$

The relationship between geographic latitude, celestial latitude and geocentric latitude is as follows:

(1) Geographical latitude eliminates the ambiguity of celestial latitude. The deviation between the geographic and celestial vertical lines is usually less than $30''$,

Fig. 3.2 Geographical and
geocentric latitudes

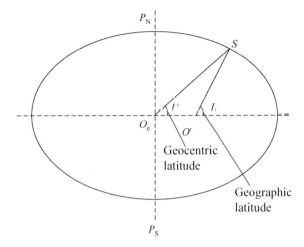

so in most applications, the geographic and celestial latitudes can be indistin-
guishable. In land navigation, it is often used as a standard representation of
latitude. Therefore, we usually call latitude the abbreviation of geographical
latitude.

(2) The relationships between geographic latitude and geocentric latitude are as
    follows: if the Earth is regarded as a sphere, then the normal will pass through the
    geocentric O_e. The geocentric latitude is consistent with the geographic latitude
    under this condition. Consider the Earth as an ellipsoid (Fig. 3.2). The geocentric
    latitude is no longer equal to the geographical latitude when the latitude is $0°$
    or $90°$, $L' = L$; at a latitude of $45°$, the maximum error is $11.5'$. Therefore,
    the maximum position deviation in the latitudinal direction is approximately
    11 nautical mile when the geographic vertical line is replaced by the geographic
    vertical line. That is, if the Earth is approximated as a sphere in navigation, the
    maximum error in latitude conversion will occur.

3.1.2 Longitude

Geographical Longitude

Longitude is defined as the central angle (or polar angle) of the Earth sphere to the
short arc between the Greenwich longitude and the longitude of a point on the equator,
symbolized as λ. The scope of longitude is $0°\sim180°$. From the Greenwich Meridian to
the East is called the east longitude, symbolled as "E"; from the Greenwich Meridian
to the west is called the west longitude, symbolled as "W". When calculating, the east
longitude is positive, and the west longitude is negative. Units are degrees, minutes
and seconds ($°, ', ''$). For example, the geographic longitude of a certain point can be
represented as $\lambda = 75° 28.2'$ E.

Fig. 3.3 Radius of earth's
primary curvature

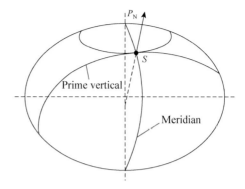

Longitude at the geographic pole cannot be defined because significant errors occur when calculating longitudes very close to the north or south poles.

Radius of the Meridian Circle Curvature

The radius of curvature moving along the north–south direction is the radius of meridian curvature R_M. It determines the rate of change in geographic latitude along a meridian. The curvature radius of the cross section of the reference ellipsoid is in the northward and downward directions of the focus (Fig. 3.3). The meridian circle is an ellipse with the major semiaxis a and the minor semiaxis b. The curvatures of each point on the meridian circle are different. The curvature is the smallest at the pole and the largest at the equator. The radius of curvature of the meridian circle R_M is given by the following formula:

$$R_M(L) = \frac{R_0(1 - e^2)}{(1 - e^2 \sin^2 L)^{3/2}} \tag{3.3}$$

R_0 is the equatorial radius, and e is eccentricity. An object moves along the meridian circle at a unit velocity, and its geographic latitude varies at a rate of change $1/R_M$.

Curvature Radius of the Prime Vertical Circle

The normal plane perpendicular to the meridian plane and intersecting the surface of the ellipsoid is also an ellipse, which is called the prime vertical circle of the point, and R_N is used to express its curvature radius. Apparently, at some point on the Earth's surface, the radius of the meridian circle curvature is not equal to that of the prime vertical circle. When the latitude of a point is zero, the prime vertical circle is the equator. Only when at the pole $L = 90°$ are R_N and R_M equal.

R_N is given by the following formula:

$$R_N(L) = \frac{R_0}{\sqrt{1 - e^2 \sin^2 L}} \tag{3.4}$$

The radius of the Meridian Curvature R_M and the curvature radius of the prime vertical circle R_N are generally referred to as the radius of the principal curvature.

3.1.3   Height

Geodetic Height

Geodetic height (h) is also called reference ellipsoid height [5, 1]. The distance is between the point and the reference ellipsoid surface along the normal line of the ellipsoid surface. The height of the vehicle is positive when it is outside the ellipsoid. It can be seen from the triangulation method that the height h_b is

$$h_b = \frac{Z_b - Z_{S(b)}}{\sin L_b} \tag{3.5}$$

Derivation is available

$$h_b = \frac{z_b}{\sin L_b} - \left(1 - e^2\right) R_M(L_b) \tag{3.6}$$

The vehicle b position can be expressed as $P_b = (L_b, \lambda_b, h_b)$. It should be noted that here, the reference coordinate system is the earth coordination system, and the projection coordination system is the local navigation system (Fig. 3.4).

The radii of the meridian circle and unitary circle are $R_M(L_b) + h_b$ and $R_N(L_b) + h_b$, respectively, when the height outside the ellipsoid is h. By the same token, the radius of curvature of the latitude circle is $[R_N(L_b) + h_b]\cos L_b$.

The velocity along the curve divided by the radius of the curve curvature is equal to the reciprocal of time of the corresponding angle of the curve. The time differential of the curve position is a linear function of the Earth's reference velocity in the local navigation coordinate system.

$$\begin{cases} \dot{L}_b = \dfrac{v_{b,N}}{R_M(L_b) + h_b} \\[2mm] \dot{\lambda}_b = \dfrac{v_{b,E}}{(R_N(L_b) + h_b)\cos L_b} \\[2mm] \dot{h}_b = -v_{b,D} \end{cases} \tag{3.7}$$

Fig. 3.4 Geodetic height diagram

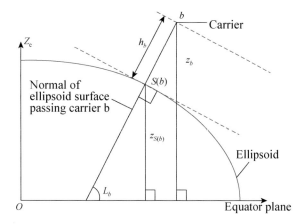

Vertical Height

Several relevant conceptual definitions about "height" [5, 1]:

(1) **Average sea level**: The average of the tide, maintaining a surface of roughly equal to gravitational potential energy.
(2) **Geoid**: A geoid is a model of the Earth's surface with constant gravitational potential energy and is a kind of equipotential surface. The geoid model is also known as the height datum, usually within 1 m above and below the average sea level.
(3) **Geoidetic height**: The height of the geoid relative to the ellipsoid. Because the Earth's gravity field varies with the region, the maximum difference between the geoid and reference ellipsoid can reach 100 m.
(4) **Terrain**: The physical surface of the earth on land. The height of a terrain is called elevation.
(5) **Orthometric height**: The height above the geoid or above the average sea level is called orthometric height or altitude.

These concepts are closely related to gravity. On the surface of the earth, the gravity vector at any point is always perpendicular to the geoid, not to the ellipsoid or the surface, although there is little difference between them.

The Relationship Between Different Heights

(1) The relationship between vertical height and geodetic height is as follows:

$$H_b \approx h_b - N(L_b, \lambda_b)$$ (3.8)

Because the geodetic height is measured along the normal line of the ellipsoid, the vertical height is measured along the normal line of the geoid (Fig. 3.5).

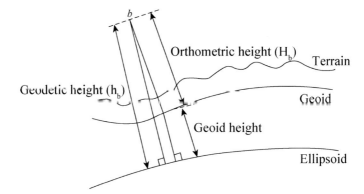

Fig. 3.5 Height of geoid, terrain and ellipsoid

(2)   In practical applications, vertical height is more valuable than geodetic height. Maps are mostly used to represent topographic features and heights relative to geoid. Vertical altitude is more accurate in aircraft approach and landing. Navigation systems often need geoid models to realize the conversion between geographic height and vertical height.

   Altitude is often measured by an altimeter, depth gauge, barometric altimeter, radar altimeter, submarine sounder, bathymeter, etc.

3.2   Navigation Parameters

In addition to position parameters, there are many kinds of navigation parameters, including speed, distance, heading, azimuth, and attitude. According to the difference in measurement, reference and vehicle, these parameters also have many different kinds, which need to be carefully distinguished.

3.2.1   Speed

Velocity is a common physical quantity. There are many different velocities in practice. Some are due to the projection of the velocity in different coordinate systems, such as the east and north velocities of the vehicle or the longitudinal and transverse velocities of the vehicle; some are due to the different references, such as the velocity relative to the ground and the velocity relative to the water (current); and some are based on the different time constants of the average calculation, such as the average velocity and the instantaneous velocity.

Speed Over the Ground and Speed Through Water

Speed is divided into two types because of the different references:

(1) Speed over the ground (SOG)

   SOG is the velocity of the vehicle along its track, i.e., the horizontal component of the velocity relative to the Earth's surface. Taking a ship as an example, it refers to the speed of a ship relative to the seabed under the influence of wind, current and wave, also known as speed made good. SOG is often used for track estimation because the estimated speed is good, and the planned speed of advance in the navigation plan is also SOG. SOG can be obtained by various means, such as the track speed output by satellite navigation receivers, Doppler logs or inertial navigation systems.

(2) Speed through water (STW)

   STW refers to the horizontal component of the ship's velocity relative to sea water, not to the surface of the earth. They are also called relative velocities in the past but are seldom used at present.

   Ship speed is the speed of a ship in still water without wind and current. The logs that can measure the ship speed relative to water under the influence of wind mainly include pit logs, electromagnetic logs and Doppler logs. The ship's velocity measured by log is called speed by log (VL). By calculating the difference between the ship's SOG and STW, the velocity of the flow field of the background environment, such as the current velocity, is obtained. In other words, the ship's SOG should be equal to the vector sum of the ship's STW and the current velocity.

Velocity Components

Different velocity components can be obtained by projecting the velocity relative to different coordinate systems. Common velocity component parameters are:

(1) Velocity components relative to geographic coordinate system

   East Velocity: The eastward projection component of vehicle velocity in the local geographic coordinate system.

   North Velocity: The northward projection component of vehicle velocity in the local geographic coordinate system.

   The east velocity and north velocity can be obtained by direct calculation of the inertial navigation system or by true heading angle conversion of the SOG by the satellite navigation system.

(2)   Velocity components relative to the body coordinate system

Longitudinal velocity: Velocity projection component along the longitudinal axis of the body coordinate system.

Transverse velocity: Velocity projection component along the lateral direction of the body coordinate system
      Longitudinal and transverse velocities can be directly output by Doppler log or by conversion.

(3)   Velocity components relative to the grid coordinate system

When navigating near the poles of the earth, it is necessary to adopt the polar grid coordinate system. The velocity can be expressed in the grid coordinate system as follows:

The grid east velocity: The projection component of the vehicle velocity in the northward direction of the grid;

The grid north velocity: The projection component of the vehicle velocity in the eastward direction of the grid.

Time-Dependent Velocity Parameters

The calculation of velocity is time dependent. According to different calculation methods, it can be divided into average velocity and instantaneous velocity.

(1)   Average velocity

The average velocity describes the average speed of an object. It is roughly expressed as the speed of an object's motion over a period of time. Average velocity is often used in route planning, mileage calculation and track calculation of vehicles.

(2)   Instantaneous velocity

The instantaneous velocity represents the speed of an object at the instant, which can also be decomposed into three-dimensional components of the body coordinate system, for example:

Heave velocity: the vertical descent of a vehicle independent of its motion attitude.

Sway velocity: the lateral motion of a vehicle independent of its motion attitude.

Surge velocity: the longitudinal motion of the vehicle independent of its motion attitude.

   Instantaneous velocity is often used in the fields of transfer alignment, ship-based helicopter takeoff and landing, which is usually provided by inertial navigation systems and inertial attitude determination systems.

Unit of Speed

Units of speed commonly used are:

(1) **Knot, kn**: International general navigation speed unit. A unit of speed expressed in nautical miles/hours, namely, 1 kn = 1 nautical mile/h.
(2) **m/s**: Speed in the most basic unit of the International Unit System, m/s, in approximate conversion. 1 kn≈0.5 m/s. Usually, used in the unit of vertical velocity.
(3) **Mach (Ma)**: The ratio of the vehicle velocity to the local sound speed. Usually, used in aircraft, rockets and other aerospace vehicles. Because the speed of sound in the air varies with different conditions, Mach is only a relative unit. Usually, 1 Ma ≈ 340.3 m/s, and the sound speed at 15 °C.

3.2.2 Distance

Sailing Distance

Sailing distance is the distance traveled by the ship within a certain time interval, and its unit is nautical mile at sea. Similar to speed, the sailing distance is divided into the following two types:

(1) **Distance over the ground (DOG)**: The sailing distance of the ship relative to the seabed under the influence of wind, current and waves, also known as distance, is good.
(2) **Distance through water (DTW)**: The sailing distance of the ship relative to the sea water. The distance between ship and sea water under the influence of wind is called distance by log (DL).

Other Distance Parameters

(1) **Depth of water** (WD): refers to the depth from the sea surface to the seabed.
(2) **Keel depth**: the depth from the keel to the seabed.
(3) **Dividing depth**: refers to the submarine depth relative to the sea surface.
(4) **Draught**: refers to the deepest length of the submerged part of a ship.
(5) **Cross-track distance** (CTD): the axial distance of the vehicle from the original orbit per unit time.

Units of Distance

(1) **Nautical mile**: the corresponding arc length of 1′ on the elliptical meridian of the earth is called 1 nautical mile.

The length of 1 nautical mile is not constant but varies slightly with latitude. The shortest length is 1842.94 m when L = 0°; the longest length is 1861.56 m when L = 90°; and it is 1852.25 m when L = 45°. Because the parameters of the earth ellipse are different, there is a slight difference in length for 1 nautical mile. For the sake of navigational convenience, a unified standard should be used. At present, most countries, including China, adopt 1 nautical mile at 1852 m, which is determined by the International Hydrographic Society and the International Conference on the Safety of Life at Sea and is the length of 1 nautical mile at L = 44° 14.0′.

(2) **Cable**: one tenth of 1 nautical mile.

A special unit for a short distance below 1 nautical mile is approximately 182 m.

(3) **Meter**: an international general unit of length. It is often used in navigation as a unit of measuring elevation and depth.

The following length units may also be used in English Chart books and materials:

$$\text{Foot(ft)} : 1\,\text{ft} = 0.3048\,\text{m}.$$

$$\text{Yard(yd)} : 1\,\text{code} = 3\,\text{ft or} = 0.9144\,\text{m}.$$

$$\text{Fathom(fm)} : 1\,\text{ft} = 6\,\text{ft} = 1.8288\,\text{m}.$$

3.2.3 Heading and Azimuth

Heading Parameters

The concept of heading is relatively broad. It is usually defined as the direction of the vehicle, that is, the angle between the heading line and a reference direction. However, there are many definitions of course and reference direction. In the field of different applications, some concepts are inconsistent. First, the common course definitions are given.

Course Line (CL): The line intersecting the bow surface of the vehicle with the local horizontal plane is called the fore-and-aft line, and the extended line from the fore-and-aft line to the bow direction is called the course line.

Definition of Direction Based on Course

Because of the difference in reference, the heading angle is divided into:

(1) **Heading (Hdg)**: Projection of the angle between the vehicle bow line and the true north (N_T) in the horizontal plane. The heading is measured by true north, which is different from the heading indicated by magnetic north (N_M) or compass north (N_C). Therefore, in many cases, the heading is also called true course (C_T).

(2) **Magnetic course (C_M)**: A course based on magnetic north is called a magnetic course.
(3) **Compass course (C_C)**: The projection of the direction indicated by the compass on the horizontal plane is called compass north. The course based on the compass north is called the gyrocompass course (C_G).

Definition of Heading Based on Course Direction of Vehicle

Aircraft and ships are affected by wind and current, which makes the heading of the vehicle inconsistent with the actual motion direction. According to the actual course of the vehicle, the following definitions are given:

(1) **Course over ground (COG)**: The angle between the course direction of the vehicle relative to the ground and true north.
(2) **Course made good (track direction)**: The angle between the true north and the track line. It is often used in marine navigation.
(3) **Yaw**: In marine navigation, yaw refers to the drift angle between the actual route and the planned route. In the aviation field, yaw refers to the angle between the specified axis of the vehicle's longitudinal axis and the local terrestrial coordinate system. When the specified axis is northward, the definition of yaw is consistent with the heading. The definition of yaw is similar to that of marine navigation when the specified axis is directed at the target direction (planned course).
(4) **Drift angle**: In the aviation field, the drift angle is the angle between the actual movement direction of the vehicle and the fore-and-aft line. In the marine field, it is also called the wind pressure deviation.

Other Direction Parameters

(1) **Wind direction**: The direction of wind is usually expressed by the 8, 16 or circumferential method.
(2) **Current direction**: The direction of current, expressed in terms of angle.

Conversion Between Different Headings

The relationship between the course line, true north (N_T), magnetic north (N_M), and compass north (N_C) is shown in Fig. 3.6. The following relations can be obtained directly:

(1)
$$C_T = C_M + Var \text{ (Magnetic deflection angle)}$$
$$= C_C + \Delta C \text{ (Compass error)}$$

(2)
$$C_M = C_T - Var \text{ (Magnetic deflection angle)}$$
$$= C_C + Dev \text{ (Magnetic compass self-deviation)}$$

Fig. 3.6 Direction
relationship

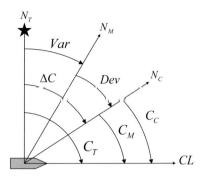

$$\begin{aligned} C_C &= C_T - \Delta C \,(\text{Compass error}) \\ &= C_M - Dev\,(\text{Magnetic compass self-deviation} \end{aligned}$$

(3)

(4) $\Delta C = Var + Dev$

Azimuth Parameters

(1) **Bearing line (BL)**: refers to the projection of the observer and the object line on the horizontal plane.
(2) **Bearing (B)**: the angle between the bearing line of the target and its reference direction (N_T, N_M, N_C, etc.) is called a bearing. Due to the different reference directions, the bearing can be divided into true bearing (B_T), magnetic bearing (B_M), and compass bearing (B_C), which are the angles between the target bearing line and N_T, N_M, and N_C (Fig. 3.7). The measurement of the bearing is rotated from their respective north to the bearing line clockwise.

(3) $C_C = C_T -$ Compass error (ΔC) $= C_M -$ Magnetic compass self-deviation (Dev)

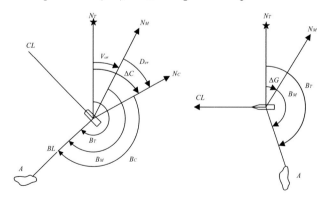

Fig. 3.7 Azimuth and its relationship

Angle Unit

Three unit systems are used to represent the direction system:

(1) **Angle**: mainly includes degree (deg, °), arc minutes (min, ′) and arc second (sec, ″).
(2) **Radian** mainly includes radians (rad) and centrad (crad). The relationship among them is 1 rad = 100 crad.
(3) **Mil**: Mil is often used in military applications. Mil is the approximate integral of milliradians, but the international definitions of mil are not the same. One circumference is approximately 6283 milliradians. China and Russia adopt the 6000-bit system; the United States and Europe adopt the 6400-bit system. The so-called 6000-bit system divides the circumference into 6000 parts, and each part is 1 mil. The notation of mil is special. The high- and low-order positions are separated by a short line, such as 1 mil is written as 0-01; 312 mils is written as 3.12; and 3000 mils is written as 30-00. In the 6000-bit system, 1bit is equal to 0.06°. If we need to convert mil to the degree, we can simply multiply the angle value by 0.06; in contrast, if we convert the degree to mil, we can divide it by 0.06 or multiply it by 16.667.

3.2.4 Attitude

In the second chapter, the definition of attitude is given based on the relationship between the geographic coordinate system and body coordinate system. The attitude parameters are further explained here.

Attitude Angle and Their Changing Rates

The pitch and roll of the vehicle are called the attitude angle and sometimes also include heading. The definitions of heading (ψ), roll (ϕ) and pitch (θ) are given in Sect. 2.2.4.4. Attitude angular velocity includes heading angular velocity, pitch angular velocity and roll angular velocity, that is, the change of ship's course, pitch angle and roll angle in unit time.

Attitude parameters are widely used. A&H plays an important role in controlling the motion state of the vehicle and ensuring the safety of navigation. For example, in the field of aviation, pilots need to know the pitch and roll of the aircraft to operate the aircraft correctly in time. A&H is mainly provided by an inertial navigation system, stabilized gyrocompass, attitude and heading reference system (AHRS) and strapdown fiber optic gyrocompass.

Deck Deformation Angle

In reality, the deck itself is not rigid but will be distorted and deformed. Different positions may face the problem of coordinate unification. The three-dimensional deformations of the vehicle can be expressed by the deformational angles in three directions, namely;

Longitudinal deflection angle: The rotation around the starboard direction of the vehicle.

Transverse deflection angle: The rotation around the for-and-after line of the vehicle.

The deck deformation angle can be measured by an inertial transfer alignment device and a deformation measurement device.

Ship Inclination

Rolling and pitching are the motions of ships under the action of water. Trim and list represent the floating state of ships in still water.

Trim: Long-term longitudinal inclination of the vehicle.

List: Long-term lateral tilt of a vehicle caused by internal forces.

Trims and lists are important for ship maneuvering. On a ship, the list can be measured by a listing instrument.

3.2.5 Sailing Parameters

Taking a ship as an example, the following are some commonly used navigation parameters.

The minimum distance of the safe point of approach (DSPA): The minimum encounter distance for two ships to pass safely considering the navigation environment and ship status.

Time to closest point of approach (TCPA): Refers to the time when the other ship arrives at the nearest meeting point of the ship.

Waypoint: A route point is the turning point of a planned route, including the starting point and the destination point.

Turning point: The point on a planned course at which the course changes.

Desired track: The route determined in the course of planning route according to the navigation order and drawn on the chart in advance.

Intended course: The direction determined in the course of the planning route.

Intended speed: The speed determined in the course of route planning.

Great Circular Route: Considering the earth as a sphere, a great circular route is the shortest course between two points on the surface of a sphere.

Thumb line: The thumb line is a curve on the Earth's surface that intersects the longitude line at the same angle.

3.3 Time Parameters

3.3.1 Time

As a basic parameter of physics, time represents the continuity of object movement and the sequence and duration of events, which contains two parts: time interval and time instant. The former describes the duration of object movement, indicating the length of time; the latter describes the exact time of object movement corresponding to the time reference instantaneously, that is, the time interval between the instant moment and the origin of time (called epochs). They are both called "time" but are different and interrelated. The main concepts of time are introduced below.

Astronomical Time

In ancient times, people had a vague concept of timing, that is, the so-called "do when sunrise and rest when sunset". According to the rising and setting of the sun from the east to the west, because of the rotation of the earth, the ancients established the unit of time, "day". In addition, the concepts of "month" and "year" are formed by observing the change relationship between the earth, the moon and the sun. In fact, the rotation of the earth and the movement between the earth, the sun and the moon are complicated. The commonly used relationships among years, months, days, hours, minutes and seconds can be listed as follows:

$$1 \text{ year} = 365.24219879 \text{ days}$$
$$1 \text{ month} = 29.530589 \text{ days}$$
$$1 \text{ day} = 23 \text{ h } 59 \text{ min } 39 \text{ s}$$

People hope to be able to use integer numbers to express the relation between a year, a month and a day. Different ancient civilizations designed different calendars according to different geographical characteristics, cultural customs, social needs, etc. Calendar is representative of the definition of human time, which is the important symbol of civilization and the foundation of social life. There are also many influential human calendars: the Chinese calendar, the Egyptian calendar, the Jewish calendar, the Muslim lunar calendar, the Mayan calendar, the Christian Julian calendar, the modern calendar and so on.

Since these units of time are obtained by observing the periodic motion of celestial phenomena (the sun, the moon and the stars), they are collectively called "astronomical time". Commonly used astronomical time s include star time, true solar time, average solar time, local time, world time (UT) and ephemeris time (ET).

Star Time

Star time is a time measurement derived from the rotation of the Earth relative to the stars. In astronomy, the average time interval that a fixed meridian circle past the equinox twice is called an average star day. An average star day is divided into 24 h, an average star hour is divided into 60 star minutes, and an average star minute is divided into 60 star seconds. That is, 1/86400 of an average star day is called a "second" of an average star time.

If we neglect the influence of the rapid fluctuation of the Earth's rotation rate and the slight change in the polar shift, the local star time is the time angle of the vernal equinox; the local true star time is the time angle of the instantaneous vernal equinox; and the local average star time is the time angle of the instantaneous mean vernal equinox. The time of stars in the Greenwich Geographic Meridian Circle is called greenwich star time. The relationship between local star time and greenwich star time can be simply expressed as.

$$\text{local startime} = \text{Greenwich star time} - \lambda \tag{3.9}$$

Here, λ is the longitude of the observer, which is positive to the west.

Solar Time

(1) **True Solar Time**

The time obtained by direct observation of the sun's motion is called the true solar time. Because the sun is a large circle, it is not easy to observe its center. Therefore, the true solar time can be obtained by calculating the time of stars passing through the heavens at midnight. Then, we obtain the true solar day, minute and second. The true solar time is convenient and intuitive, but due to the uneven rotation of the earth, the true solar day length varies. For example, there is a difference of 51 s in the length of the two true solar days between September 16th and December 12th each year. Therefore, the true solar time cannot be used as a unit of time measurement.

(2) **Mean Solar Time**

In 1820, the French Academy of Sciences defined 1/86400 of the average daily length of the true sun (average solar day) throughout the year as 1 s. At that time, people thought that the average position of sun was fixed and unchanged, so it was called the "mean solar day". Later, Newcon proposed replacing the real sun with an imaginary sun, not moving in the solar ecliptic but moving uniformly on the equator but at the same speed as the real sun moving uniformly on the ecliptic and as close as possible to the real sun, thus eliminating the influence of the elliptical orbit motion of the Earth's revolution and the ecliptic obliquity. The moment that the sun was at the bottom of midnight on January 1st, 1962, is considered the start epoch of the day.

Local Time and Time Zone

There are 24 time zones in the world. Each time zone covers the area in which the longitudinal interval is 15°. The zero zone begins with the meridian and starts with the western meridian 7.5° to the eastern meridian 7.5°. Beijing longitude is 116° 19′ E, and the real local Beijing time is approximately 15 min later than the official Beijing time (120° E).

World Time

(1) **Universal Time (UT)**: The mean solar time observed at the Greenwich Observatory in the United Kingdom from midnight is called Universal Time (UT), also referred to as Greenwich Mean Time (GMT). UT is widely used in international bulletins and the literature in terms of time and frequency.

(2) **Universal Time (UT0)**: In 1956, astronomy stipulated that UT0 is the world time obtained directly from the observation of stars without any modification (except for the modification when the star time is changed into the mean solar time).

(3) **Universal Time (UT1)**: Due to the influence of solar wind and tides, the rotation of the earth is not uniform. The variation in the rotation rate of the earth includes long-term, seasonal and irregular changes. Slow in the first half year and fast in the second half year, the polar axis also changes very slightly. These effects make the UT0 system inaccurate and have some errors. If UT0 is modified by the pole shift, the world time obtained by the pole shift correction is recorded as UT1; then,

$$UT1 = UT0 + \Delta\lambda \tag{3.10}$$

Among them, $\Delta\lambda$ is a correction of the average polar shift.

(4) **Universal Time (UT2)**: If the periodic variation of the Earth's rotation rate is corrected for UT1, UT2 will be obtained, that is:

$$UT2 = UT1 + \Delta T_s = UT0 + \Delta\lambda + \Delta T_s \tag{3.11}$$

where ΔT_S is a revised value of the periodic variation in the Earth's rotation rate. UT2 systems also have an error of $\pm 1 \times 10^{-8}$ s. It is estimated that a cumulative error of nearly 2 h will occur after two thousand years.

Ephemeris Time (ET)

To meet the demand for precise time, in 1958, the International Astronomical Union (IAU) decided to replace world time with ephemeris time (ET) from 1960. The ephemeris day is defined as the result of a tropical year divided by 365.24219879, and the ephemeris second is the result of an ephemeris day divided by 86,400. After hundreds of years of observations, astronomers have found that although the Earth's revolution speed is not constant, its rotation period is quite stable. Therefore, it can be used to determine the basic unit of time, i.e., second. Although the ephemeris time is defined by the sun, it is difficult to observe the sun. In reality, it is determined by observing the Moon to obtain the difference between the ephemeris time and the world time $\Delta T = ET - UT$ and then obtain the ephemeris time by conversion.

Atomic Time (TA)

The irregularity of the shape and margin of the moon, together with the deceleration of the moon caused by the Earth-Moon tides, will affect the accurate measurement of ΔT. Further considering the accuracy of the celestial instruments, the accuracy of the ephemeris time that can be obtained by this method is 10^{-9} magnitude. This is still unable to meet the needs of high accuracy of modern science and technology. The emergence of atomic time has made the time measurement transition from the macrocosmic world of astronomy to the microscopic world of quantum physics. The accuracy of time measurement has also been improved by several orders of magnitude.

The study of quantum physics tells us that the electrons around the nucleus can produce energy level transitions. When high-level electrons transit from high-level to low-level, they emit a certain amount of energy, i.e., electromagnetic waves with stable frequencies. The frequency of the electromagnetic wave radiated by the transition depends on the physical properties of the atom itself and the energy provided by the outside world. The frequency standard is the atomic frequency standard, which takes the frequency radiated by the atom when it transitions from high to low energy levels. The time measurement system based on the atomic frequency standard is called atomic time, abbreviated as *TA*.

Atomic time is a time reference generated by a continuously running atomic clock. However, the atomic time established by the atomic clocks could be different because their starting points are different. In addition, even if the same starting point is chosen, because of the difference in frequency accuracy and stability of the clocks, the shown time value after long-term accumulation will be significantly different. For this reason, in the initial stage of the establishment of *TA*, in addition to the common prescribed starting point, the mean atomic time is derived by using the statistical averaging method of multiple clocks to obtain as accurate and uniform a time reference as possible.

(1) **Atomic Time Second**

In the 1960s, after the development of the cesium beam atomic frequency standard, people measured the cesium beam atomic frequency standard (1955–1958) with the seconds of ET and obtained the oscillation number of the cesium beam resonator during the ET-second period as follows:

$$f_{cs} = (9192631770 \pm 20)\,\text{Hz} \qquad (3.12)$$

The 13th International Conference on Metrology in 1967 defined atomic time-second as the duration of 9192631770 periods of radiative oscillation of transition radiation between two hyperfine levels of cesium atomic ground state in a zero magnetic field at sea level. At the same time, it is stipulated that the conversion relations of second, minute, hour, day, month and year of atomic time are still the same as those of world time.

(2) **Local Atomic Time (TA (k))**

The atomic time derived from a clock group consisting of several atomic clocks in several laboratories in a region is called local atomic time. According to the definition, atomic time can be provided by any atomic clock that runs continuously after the initial time has been determined. Each laboratory can use the following methods to derive their local atomic time: (1) continuous large cesium clocks; (2) regular calibration of continuous rubidium clocks, commercial small cesium clocks or hydrogen clocks with large cesium clocks; (3) a combination of several commercial small cesium clocks, etc.

(3) **International Atomic Time (TAI)**

International atomic time (TAI) is a time scale established and maintained by the International Authority on the basis of a large number of atomic clocks operating in time-keeping laboratories around the world. The TAI started at 0:00 on January 1, 1958, which was very close to the UT, only 0.0039 s different. Because of the inhomogeneity of the Earth's rotation speed, the UT has been approximately one second slower than the atomic time in the past two decades. After determining the initial time of the TA, the difference between the two has accumulated year by year and reached 32 s in 2001.

Coordinated Universal Time (UTC)

AT with high stability meets the needs of time uniformity and accuracy for users with high precision time and frequency, but most users are still accustomed to UT, which is closely related to people's lives, and it has practical value in geodesy, celestial navigation, tracking and measurement of space flight objects and other fields [6]. It is impossible for time service departments to transmit a set of time signals to meet two different needs, so the problem of coordination between the TA and UT arises.

In 1970, the International Astronomical Society (IAU) proposed that UTC became the international standard time since January 1972. It represents the combination of two time scales TAI and UT1, defined by the following two formulas:

$$UTC(t) - TAI(t) = n \text{ seconds (n is integer)} \qquad (3.13)$$

$$|UTC(t) - UT1(t)| < x \text{ second } (x < 1) \qquad (3.14)$$

Before 1974, $x = 0.7$, and after that, it changes to 0.9.

Make UTC seconds exactly equal to TAI seconds, and make UTC close to UT1 in time value. The difference between UT1 and TAI caused by the uneven rotation rate of the earth is compensated by adding or subtracting leap seconds in UTC time. Leap seconds are scheduled and adjusted on 30 June or 31 December, i.e., The last minute of UTC on 30 June or 31 December is 61 s or 59 s.

There have been 27 leap seconds since 1972. The last leap seconds were at 7:59:59 on January 1, 2017 in Beijing time. Leap seconds occur every few years. Leap seconds have little impact on the lives of ordinary people, but they have a great impact on space, communications, finance, military and other fields, which have strict requirements for the accuracy of time continuity.

It is worth noting that due to the uneven rotation of the Earth's long-term slow-down, the difference between UT1 and IAT is increasing. It is estimated that UT1 will be half a day behind IAT in 4000 years. If UTC is used all the time, it may reach leap seconds twice a month.

3.3.2 Frequency

Definition of Frequency

In physics, the number of vibrations per unit time is called the frequency of vibration. Frequency and period are both physical parameters indicating the speed of vibration.

Measurement of Frequency

The 7th Research Group of the International Telecommunications Union (ITU) recommended a vocabulary of time and frequency in 2001. Only three commonly used terms are listed here to describe the performance of time–frequency standard parameters.

Correction: Represents the approximation of a measured physical quantity to its true value.

Accuracy: This represents the degree of coincidence between a series of individual measurements. Accuracy is often expressed by the standard deviation of measurements.

Frequency stability: Frequency stability represents the frequency variation of signals in a given time interval.

Appendix 3.1: Questions

1. What is the National Comprehensive PNT system? How about its great influence? What are the PNT parameters?
2. How can we understand the relationship between positioning and navigation? What are the common positioning parameters? What are the latitude definitions? What latitude is commonly used in inertial navigation?
3. What are the common altitude parameters? What are the differences between geodetic height, orthometric height, elevation, ellipsoidal height and sea level height?
4. What are the types of speed parameters? Can you explain the definition, characteristics and applications of different speeds?
5. What are the types of heading and bearing parameters? Can you explain the definition, characteristics and application occasions of these different parameters?
6. There have been different calendars in the world, such as the Chinese calendar, Egyptian calendar, Jewish calendar, Muslim lunar calendar, Mayan calendar, and Christian Julian calendar. Thinking about the causes of these different calendars. Can we list the time units adopted in ancient civilizations such as China, India and Maya and understand the commonalities and differences in the understanding of time of different human civilizations?
7. Please point out the types and differences between astronomical time and atomic time. What is coordinated universal time? What is leap second? Can you tell when the world's last leap second and the next leap second?
8. Understand the importance of terminology for professional learning. How can we dialectically understand the relativity of technology development and concept accuracy and consistency? There are many famous expressions of the inherent contradictions of concepts in traditional Chinese classics. Considering the PNT parameters definition, deepen the understanding of the definition and change of the concept.

Appendix 3.2: The Importance of Understanding the Concepts

Concepts are the basis of scientific research and logic reasoning. Misunderstanding of concepts may present problems in communication, and hinder the progress of scientific research and other related work. In fact, to enhance scientific literacy and deepen professional understanding calls for better awareness of accurate understanding and further exploration of academic concepts.

Professional learning expects the habit of exploring the definition of technical terms thoroughly. Proper use of technical term, an important indicator of expertise, stems from the accurate understanding of terminology and concepts. For example, the term "latitude" may refer to "geocentric latitude", "geographical latitude", or "celestial latitude". What is the difference? What is the connection? What are their unique functions? Without systematic learning of navigation, one can be easily confused with these concepts. Only by understanding these terms can we truly understand the concept of "latitude" and then establish an enhanced professional understanding.

It is also worth noticing that the accuracy of a concept is also relative from a dialectical viewpoint. Because the continual development of science and technology would generate new technical terms and concepts, the concepts themselves are also evolving. So far, many terms and concepts have not been clearly defined yet, which has caused many practical problems. Therefore, we should learn the connotation and essence of a concept with an open mind. When promoting a sense of professionalism, we should also make our own contributions to the definition of concepts and the development of the profession.

Chapter 4
Direct Positioning

Positioning is the ability to precisely determine one's location in two dimensions (or three dimensions) referenced to a standard geodetic system (such as WGS84). The basic principles of navigation can be summarized and described in a simple way.

Let us start with an example of everyday life. In cities, when people walk to their destination, they must first know the position of the destination, then walk for a while and a certain distance, and then identify some obvious signs such as landmarks, confirm the next direction and walking time, and adjust the route. To keep repeating this process until reaching the destination.

During this process, people unconsciously use two different ways of navigation. One is to rely on external objects to judge their own place and target orientation and to change the route accordingly. Another is to judge the travel distance only by estimating its speed and time of movement. The difference between these two modes is whether they depend on external references. In navigation, we can call the first mode direct positioning or navigation based on an external reference, which is the most widely used positioning mode. The second mode is dead reckoning, which is an autonomous positioning method based on the information of the vehicle itself. This chapter introduces the direct positioning method based on an external reference.

4.1 Basic Principles of Direct Positioning

4.1.1 Citation

Direct positioning is the most common positioning method of navigation. It mainly determines the position of the vehicle itself by measuring the geometric relationship between the vehicle and the known reference point. The position can be either on a plane or sphere in two dimensions or three dimensions. The geometric relationship

H. Bian et al., *Essentials of Navigation*, https://doi.org/10.1007/978-981-99-5636-4_4

Fig. 4.1 Azimuth position line

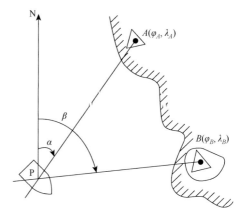

mainly refers to the angle or distance that can be measured between the vehicle and the external references.

Examples of the direct positioning method are given below.

Example 1 The direct positioning method based on angle measurement.

The vehicle position is determined based on the measured azimuth of two external references. The intersection of two equiaxed lines passing through two external reference points is the vehicle position (Fig. 4.1).

Example 2 The direct positioning method based on distance measurement.

The vehicle position is determined based on the distance of two different benchmarks, which are measured simultaneously in a plane. In Fig. 4.2, the two benchmarks are considered as the center of the position circle, the radius of which is the measured distance. Two position circles can create two intersections, which are close to the estimated vehicle position.

The above examples show that this direct positioning method needs at least two known measured angles or distances to obtain the position of the vehicle.

4.1.2 General Formula for Direct Positioning

Direct positioning is widely adopted in various systems, such as GNSS, radio positioning, celestial positioning, geographic positioning, and underwater acoustic positioning.

Suppose $\mathbf{r} = (x_{\beta\alpha}^{\gamma}, y_{\beta\alpha}^{\gamma}, z_{\beta\alpha}^{\gamma})$ is the projection of the Cartesian position vector of the origin of body coordinate system b relative to the origin of coordinate system β in the coordinate system r. $\mathbf{r}_{\beta t_i}^{\gamma} = (x_{\beta t_i}^{\gamma}, y_{\beta t_i}^{\gamma}, z_{\beta t_i}^{\gamma})(i = 1, 2, \cdots, N)$ is the projection of the position vector of the known external reference points t_i relative to the origin of coordinate system β in coordinate system r. $h_{pi}(\mathbf{r}_{\beta\alpha}^{\gamma}\mathbf{r}_{\beta t_i}^{\gamma})$ is the position function

Fig. 4.2 Distance position
line

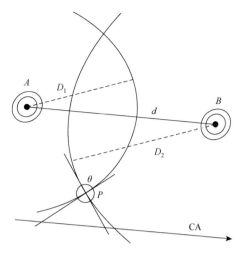

that describes the geometric relationships between $r^{\gamma}_{\beta t_i}$ and $r^{\gamma}_{\beta \alpha}$. Considering the
measurement error w_{pi}, the actual observed value z_{pi} can be described as:

$$z_{pi} = h_{pi}(r^{\gamma}_{\beta b}, r^{\gamma}_{\beta t_i}) + w_{pi} \quad (i = 0, 1, 2 \cdots, N) \tag{4.1}$$

The measurement equations for multiple measurements are obtained as follows:

$$z_p = h_p(r^{\gamma}_{\beta b}, R_p) + w_p \tag{4.2}$$

where $z_p = \begin{bmatrix} z_{p1} \\ z_{p2} \\ \vdots \\ z_{pm} \end{bmatrix}$, $h_p = \begin{bmatrix} h_{p1} \\ h_{p2} \\ \vdots \\ h_{pm} \end{bmatrix}$, $R_p = \begin{bmatrix} r^{\gamma}_{\beta t_1} \\ r^{\gamma}_{\beta t_2} \\ \vdots \\ r^{\gamma}_{\beta t_m} \end{bmatrix}$, $w_p = \begin{bmatrix} w_{p1} \\ w_{p2} \\ \vdots \\ w_{pm} \end{bmatrix}$

m is the number of observations.

The position can be solved from the above equations by the least squares method.

$$r^{\gamma}_{\beta b} = h_p^{-1}(z_p, R_p, w_p) \tag{4.3}$$

where $h_p^{-1}()$ is the inverse function of $h_p()$.

Formulas (4.1) to (4.3) are direct positioning formulas for navigation.

Note: $h_p()$ can be a unified positioning function or a different positioning function.

4.1.3 Position Function

Type of Position Function

The determination of the position function is the key problem of direct positioning.

Position function definition: Observing the carrier and the external reference point can obtain observed value that characterizes the specific relative geometric relationship between the two. The mathematical expression of the set of possible position points of the carrier satisfying the same observation value in the specified space is called the position function of the carrier.

It can be expressed as a position isoline function or a position isosurface function. The isoline describing the possible position geometric trajectory of the carrier is called the position line of the carrier, and the analytical expression of the position line in the given coordinate system is called the position line function. The possible position geometry surface of the carrier can be described by the isosurface, which is called the position surface of the carrier, and the analytical expression of the position surface in the given coordinate system is called the position surface function. The position line function and the position surface function are collectively called the position geometry function.

Typical position geometric functions can be divided into two categories: "angle-dependent" and "distance-dependent". The angle-dependent position function can be further divided into two different cases: coplanar and noncoplanar of the reference point and observer; it can also be divided into two different cases: measurement reference and no measurement reference. The measurement reference method needs to establish azimuth or horizontal reference, and the no measurement reference method mainly depends on the angle relationship between external reference points.

The distance-dependent position function can be divided into many cases, such as the distance between the reference and observer, the distance difference or the distance sum between two references. Different direct positioning methods need to establish different position geometry functions.

Positioning Solution Based on Position Function

When the carrier position is obtained by the direct positioning method, either the same position function or different position functions can be adopted according to the various measurement methods used. According to the different combinations of various position functions, various positioning methods can be obtained. The direct positioning method can not only achieve positioning by a single navigation mode but also support comprehensive positioning by various navigation modes. When it is uncapable to achieve carrier positioning by a single method, a variety of different methods and types of position lines can be used to calculate the carrier position solution, such as combination positioning by azimuth and distance, combination positioning by azimuth and horizontal angle positioning, combination positioning by azimuth (or distance) and isobath positioning.

The method of solving the position solution by using the position function is very flexible. For example, set up the analytic equations of several position functions to solve the resolution. When there are nonlinear equations or multiple redundant equations, we should consider and analyze the specific mathematical model of measurement error. The determination of ship position can also be realized by the classical method of geometric calculation. When the increase in the position function causes an increase in the intersection point, the intersection is no longer a point but a region. At this time, error processing analysis can increase the positioning accuracy.

Features of Direct Positioning

Compared with the system based on dead reckoning positioning, positioning systems based on the direct positioning method have the following features:

(1) Providing absolute positioning results

 Because the precise positions of the external reference points in the navigation coordinate system (such as inertial coordination system or the earth coordination system) are known, the absolute position in the navigation coordinate system can be calculated directly, such as geographic longitude, latitude, height and other information in the earth coordination system.

(2) Achieving fast positioning

 The real-time spatial geometry information between the observer and the reference is achieved by different measurement methods from direct positioning systems. At present, the speed of computers is relatively very fast, and the time consumed in data processing can be neglected. The time loss mostly depends on the signal identification, acquisition and tracking, which mainly affects the first positioning time. If the first positioning time of direct positioning is shorter, normal positioning is usually real-time and faster.

(3) Accurate positioning without accumulative error

 The error of direct positioning is not accumulated because of the real-time acquisition of external reference information for the solution. Compared with any dead reckoning positioning, it can usually achieve high positioning accuracy.

(4) Positioning error is related to the spatial geometric configuration of references

Multiple sets of external reference points can often be selected for direct positioning. The error is related not only to the accuracy of the angle, distance of measurement and time synchronization but also to the spatial geometric relationship of external references relative to the observer. Figure 4.3 takes the two-dimensional positioning case of the simplest bidirectional distance measurement as an example. The arc in the figure represents the mean and error interval of each distance measurement; the shaded part represents the uncertain range of the localization solution; the arrow direction represents the sightline of the vector from the observer to the reference signal transmitter. The positioning error is minimal when the sightlines

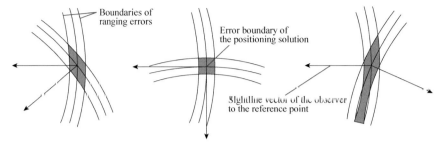

Fig. 4.3 Effect of the spatial geometric configuration of references on positioning accuracy in two-dimensional ranging measurement

of the vectors are perpendicular to each other under the condition of a given range measurement accuracy. Uncertainty of positioning or standard deviation of positioning error is associated with uncertainty of measurement or the standard deviation of measurement error through dilution of precision (DOP) [7].

4.2   Terrestrial Positioning

Terrestrial positioning is the earliest direct positioning application, such as prehistoric triangle positioning in ancient England or various terrestrial positioning methods using navigation lighthouse and navigation aids. These methods are very versatile, and many wireless indoor positioning technologies today also adopt the same positioning principle as terrestrial positioning.

4.2.1   Multipoint Positioning Based on Angle Measurements

Positioning Method Based on Two Azimuths on a Plane

Positioning based on azimuth, also known as cross positioning based on different azimuth measurements, refers to the method of simultaneously observing orientations to two or more known reference points to determine the carrier position (Fig. 4.4). The azimuth-based positioning method is simple and rapid. It is a common method of navigation.

The position function of this method belongs to the position function relative to angle measurement, and the reference and the observer are coplanar. Such angular observations are usually based on the true north, and the angular measurements are the azimuth of the reference relative to the observer. In most applications of small-range navigation and positioning, the external reference and the observer can be regarded as being in the same plane, and then the position line (also known as the azimuth position

Fig. 4.4 Positioning based
on two azimuth on a plane

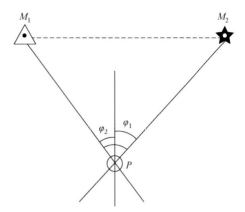

line) in which the observer is located is a ray (Fig. 4.5). However, in most near-ground carrier navigation, the reference point and the observer cannot be regarded as being in the same plane but as being in the same sphere (or ellipsoid). There is no straight line on the sphere, where the observer's position line is an arc.

If the measurement is the relative azimuth ψ_i between the carrier $r_{\beta\alpha}^{\gamma}$ and the known reference $r_{\beta t_i}^{\gamma}$, according to formula (4.3), the analytic expression of the azimuth position function of the carrier on the plane can be expressed as:

$$h_{pi}\left(r_{\beta\alpha}^{\gamma}, r_{\beta t_i}^{\gamma}\right) = \psi_i = \arctan \frac{y_{\beta\alpha}^{\gamma} - y_{\beta t_i}^{\gamma}}{x_{\beta\alpha}^{\gamma} - x_{\beta t_i}^{\gamma}}, i = 1, 2, \cdots, N \qquad (4.4)$$

Because the positioning of two azimuths on a plane cannot judge whether there are errors in the observation and the accuracy of the positioning result, the three azimuth-based positioning method is more commonly used in practice. The three azimuth-based positioning methods can obtain three azimuth position lines at the same time by observing the three reference points simultaneously. The intersection point of the position lines is the carrier position at the obtained observation time.

Fig. 4.5 Angle-dependent
position lines

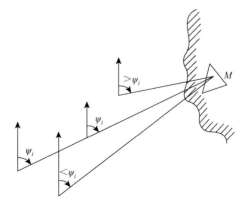

Plane Positioning Based on Two Horizontal Angles

The method of determining the carrier position by simultaneously observing the two horizontal angles between the three references is called the three targets and two horizontal angles positioning. This method is a common method in the field of navigation. The horizontal angle can be obtained by measuring the horizontal angle of two external references directly relative to the observer by a sextant. Concrete implementation can be realized by using a three-bar locator, drawing method and so on. Unlike the previous method, such methods may not require true north reference.

   If the external reference and the observer are in the same plane, the carrier position line of this method is an arc with the horizontal angle as the circumference angle (Fig. 4.6). From the geometric theorem, we can see that the horizontal angle between the two external reference points is the same at any point on the arc. The horizontal angle position line is part of the outer circular arc of the triangle connected by the carrier and the two reference points.

   If the observation is the horizontal angle ϑ between two known references $\boldsymbol{r}^{\gamma}_{\beta t_1}$ and $\boldsymbol{r}^{\gamma}_{\beta t_2}$ to the observer $\boldsymbol{r}^{\gamma}_{\beta\alpha}$, the analytic expression of its position function can be expressed as:

$$h_{pi}\left(\boldsymbol{r}^{\gamma}_{\beta\alpha}, \boldsymbol{r}^{\gamma}_{\beta t_1}, \boldsymbol{r}^{\gamma}_{\beta t_2}\right) = \vartheta = \arctan\frac{y^{\gamma}_{\beta\alpha} - y^{\gamma}_{\beta t_1}}{x^{\gamma}_{\beta\alpha} - x^{\gamma}_{\beta t_1}} - \arctan\frac{y^{\gamma}_{\beta\alpha} - y^{\gamma}_{\beta t_2}}{x^{\gamma}_{\beta\alpha} - x^{\gamma}_{\beta t_2}} \qquad (4.5)$$

Fig. 4.6 Horizontal angle-based position line

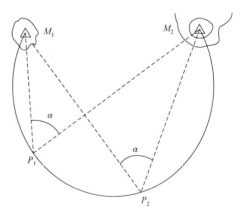

4.2.2 Multipoint Positioning Based on Distance Measurements

The distances D_1 and D_2 from the carrier to the reference M_1 and M_2 are measured (Fig. 4.7). The circular arc is drawn with the reference points M_1 and M_2 as the center and the observation distances D_1 and D_2 as the radius. The two position lines usually have two points of intersection. The one close to the calculated position of the carrier is the observed position P of the carrier at that time. This method is called the two-distance positioning method of plane.

In fact, the position line is always a circle whether the known reference and the observer are on the same plane or on the same sphere (Fig. 4.8). If the measurement is the distance between the carrier $r_{\beta\alpha}^{\gamma}$ and the known reference $r_{\beta t_i}^{\gamma}$, the analytic expression of the two-dimensional position function can be expressed as:

$$h_{pi}\left(r_{\beta\alpha}^{\gamma}, r_{\beta t_i}^{\gamma}\right) = d_i = \sqrt{\left(x_{\beta\alpha}^{\gamma} - x_{\beta t_i}^{\gamma}\right)^2 + \left(y_{\beta\alpha}^{\gamma} - y_{\beta t_i}^{\gamma}\right)^2}, i = 1, 2, \cdots, N \quad (4.6)$$

Fig. 4.7 Two distance-based positioning line on a plane

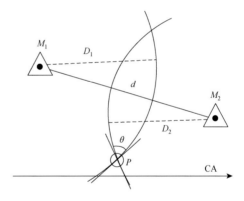

Fig. 4.8 Distance circle position line on a plane

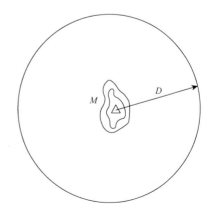

4.2.3 Triangular Positioning

Triangular positioning is often used in navigation and surveys. Triangular positioning, also known as triangulation, is widely used. It has many definitions and is not uniform. Simply to say, as long as according to the triangle principle, given two vertex positions of multiple triangles and related geometric elements, determining the third point position method can be called the triangulation method.

As long as the length of the three sides of a triangle is fixed, the shape and size of the triangle are completely determined. This is an actual coplanar two-distance positioning method. The third point can also be found by knowing the length of one side and the angles between that side and the other two sides. For example, two or more detectors are used to detect the orientation of the target at different positions, and then the position and distance of the target are determined by using the principle of triangular geometry. For example, the location of the two reference points is given, and the orientation at which each reference point reaches the observer is measured. The observer's position can also be estimated. This is an actual coplanar two-azimuth positioning method.

4.2.4 Running Fix

Multipoint positioning or positioning based on distance and azimuth from one single reference is achieved by the intersection of two or more position lines at the same time. In many cases, angles are easy to observe, but distances are difficult to obtain. If there is only one reference within the horizon and only one angle of the reference's azimuth can be obtained, then only one position line based on the reference point can be obtained at one moment, so the carrier position cannot be determined. The running fix method is used to solve this problem in navigation.

The running fix method converts two position lines of one reference measured at different times to two position lines at same time to make them intersect with each other to obtain the ship position. This is an important way to find ship positions in navigation. As shown in Fig. 4.9, let the ship sail in a straight line. The azimuth of the external reference M is measured at time t_1, and the azimuth position line A_1M is obtained; if the measurement error is not taken into account, the ship at time t_1 should be located on A_1M, that is, it may be located at any point on the line such as a, b, c and d.

After that, the ship continued its voyage along a straight line with a distance S. At time t_2, the ship may accordingly be located at any point on the line $A_1'M'$ such as a', b', c' and d'. The straight line $A_1'M'$ is the azimuth line A_1M, which moves the distance S along the track, and the resulting ship line from time t_1 to time t_2 is called the transfer position line. Obviously, at time t_2, the ship should be on the transfer position line $A_1'M'$. To measure the azimuth of the external reference again, another

Fig. 4.9 Schematic diagram
of running-fix positioning

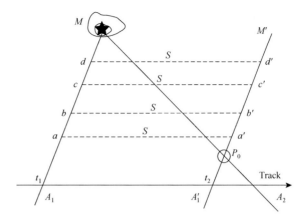

position line A_2M is obtained. The point of intersection of A_2M and $A_1'M'$ is the
position of the ship at moment t_2.

The Running fix method makes full use of the speed of the ship, which is a
combination of the geometric direct positioning mode based on the external reference
and the dead reckoning mode based on the ship itself. The running fix method is
suitable for all kinds of position lines, for example, in terrestrial positioning, the
azimuth position line, distance position line, etc., all can be used. Solar positioning
in celestial positioning is also a kind of running fix method, that is, two solar ship
position lines measured at different times are transferred to the same time to make
intersections to obtain the ship position. This method can be used in oceans where
no external reference can be observed.

4.3 Celestial Positioning Principle

4.3.1 Basic Principles of Celestial Positioning

In celestial navigation, the position of the carrier can be determined by using optical
instruments to observe the altitude angle and azimuth angle of the star [8]. The so-
called altitude angle is the angle between the sightline of the star and the local horizon
(Fig. 4.10). Because the North Star is basically on the Earth's axis of rotation and
is very far away from the Earth, the sightlines of any point on Earth to the North
Star can be considered parallel to each other. Suppose that the observer is located at
point P, and the altitude angle of the North star observed from point P is h. It can be
seen that the geographic latitude of P is equal to h. In this way, the position of the
carrier can be determined on a circle of equal altitude. Using the measured altitude
angle of two stars, two circles of equal altitude can be determined, and one of the
two intersection points of the two circles is the carrier position.

Fig. 4.10 Circle of equal
altitude

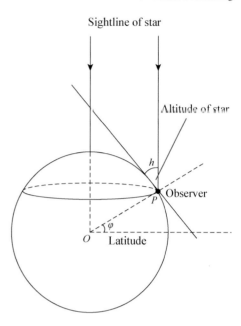

Let A and B be two celestial bodies, as shown in Fig. 4.11. The intersection points between the lines of celestial A and B and geocentric O and the surface of the earth are a and b, which are called the substellar points. If the geographical location of the substellar points can be known, the distances from the observer to the substellar points are z_a and z_b, respectively. Then, taking a and b as the centers of the circle of equal altitude, taking z_a and z_b as the radii, and making two arcs near the calculated carrier position, the intersection point M is the celestial position point.

However, the external reference point used for celestial positioning is the celestial body, whose substellar point on the ground is invisible; because of the rotation of the earth, the celestial body shows a constant apparent rotation motion relative to the earth, and its substellar point position changes with time.

Therefore, the following needs to be solved: the location of the substellar point of the celestial body to determine the center of the circle of equal altitude; the distance from the observer to the substellar point of the celestial body to determine the radius of the circle of equal altitude of the observer; and the position of the observer can be plotted on the chart to obtain its latitude and longitude.

4.3.2 Position Function Based on Altitude Angle

The celestial positioning method is actually an angular-dependent positioning method under the condition that the reference and the observer are not coplanar. At this point, it is necessary to measure the altitude angle of the external reference

Fig. 4.11 Celestial positioning

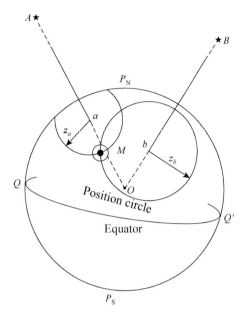

(such as a star) relative to the horizon at which the carrier is located, so it is necessary to obtain the local horizon of the observer as the observation reference.

If the external reference can be regarded as at an infinitely far position, the set of all possible position points of the observer in the 3D space is located on a cylinder, that is, the surface of the carrier position is a cylinder. If the reference point cannot be regarded as at an infinitely far position and can only be regarded as a point in space, the position surface of the observer is a cone. Both a cylinder or a cone intersects a plane or a sphere is still a circle (as shown in Fig. 4.10).

At the same time, the surface of the observer position can be either on a plane or a sphere. For example, some small-range navigation and positioning can be regarded as positioning problems on a plane; in large-range navigation and positioning, because the earth cannot be approximated as a plane, the positioning and navigation problems at this time must be regarded as spherical (even ellipsoid) positioning and navigation problems. For navigation and positioning in a plane, if the reference and the observer are not in the same plane, the position surface of the observer intersects with the plane in which the observer is located in a circle. For spherical navigation and positioning, if the reference and the observer are not in the same sphere, the intersection of the position surface of the observer and the sphere the observer located is still a circle.

In celestial navigation, the observation of stars other than the sun or moon approximates the reference as far infinity, and the circle obtained by its position surface intersected with the Earth's surface is called a circle of equal altitude.

The position function based on the altitude angle can be expressed as follows:

If the observation is the angle η (i.e., the altitude angle) between the sightline vector $r^{\gamma}_{\beta t_i}$ of the observer to the reference and the plane with the normal unit vector $n = (x_n, y_n, z_n)$, the analytic expression of the position function is:

$$h_{pi}\left(\boldsymbol{r}_{\beta\alpha}^{\gamma}, \boldsymbol{r}_{\beta t_i}^{\gamma}, \boldsymbol{n}\right) = \eta = 90° - \vartheta, i = 1, 2, \ldots N \tag{4.7}$$

where ϑ is the angle between the line-of-sight vector $\boldsymbol{l} = \boldsymbol{r}_{\beta\alpha}^{\gamma} - \boldsymbol{r}_{\beta t_i}^{\gamma}$ and normal unit vector of the plane, satisfying

$$\cos\vartheta = \frac{\boldsymbol{l} \cdot \boldsymbol{n}}{|\boldsymbol{l}||\boldsymbol{n}|} = \frac{x_n\left(x - x_{t_i}\right) + y_n\left(y - y_{t_i}\right) + z_n\left(z - z_{t_i}\right)}{\sqrt{\left(x - x_{t_i}\right)^2 + \left(y - y_{t_i}\right)^2 + \left(z - z_{t_i}\right)^2}} \tag{4.8}$$

and

$$x_n^2 + y_n^2 + z_n^2 = 1 \tag{4.9}$$

.

4.3.3 Celestial Positioning Based on Altitude Intercept

Finding the Position of the Substellar Point of the Celestial Body.

To determine the geographical coordinates of observers, the relationship between celestial body coordinates and geographical coordinates is studied. The relationship between the earth and the celestial sphere is shown in Fig. 4.12.

In Fig. 4.12, the inner sphere represents the earth. The outer sphere is an imaginary sphere with the earth as its center and an infinite radius. It is called the celestial sphere. Points, lines and circles on the celestial sphere are the extensions of points, lines and circles on the earth. For example, the equator, the Arctic and the Antarctic of the Earth extend to the celestial sphere, which corresponds to the celestial equator, the celestial Arctic and the celestial Antarctic, so the celestial coordinates of celestial bodies correspond to the coordinates of their projection points on the Earth. where

Fig. 4.12 Relations between the celestial sphere and the earth

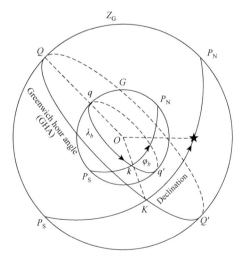

celestial point B on the celestial sphere is represented by the Greenwich hour angle (GHA) and declination, which are equal to the longitude λ_b and latitude φ_b of the projection point, respectively. Namely,

$$\text{position of celestial object B} \begin{cases} \text{GHA} = \text{longitude} \\ \text{declination} = \text{latitude} \end{cases} \text{position of subpoint } b$$

The values of the Greenwich hour angle and declination of celestial bodies can be obtained from the nautical almanac according to the world time when the celestial body is observed, so the latitude and longitude of the substellar points of the observed celestial bodies can be converted and obtained. Therefore, the substellar points can be marked on the chart.

Calculating the Distance from the Ship to the Substellar Point

As shown in Fig. 4.13, B is a celestial body, b is the substellar point of celestial body B on the earth, and M is the observer (ship). The distance from the observer to the substellar point is $M_b = Z_b$. Since the celestial body is far from the earth, the light from the celestial body to the earth's surface and earth center can be regarded as parallel light. $M_b = Z_b = 90° - h$, where h is the altitude of the celestial body. It can be obtained by observing the angle of the celestial body to the horizon on the sea with a sextant and making some correction.

According to the world time when observing celestial bodies A and B, looking up to the nautical almanac, the Greenwich hour angle and declination of two celestial bodies are determined to obtain the longitude and latitude of substellar points a and b. Taking a, b as the center and the distance z_a, z_b as the radius to make two arcs close to the calculated position, the intersection point M is the ship's position determined by celestial observation.

Because the radius of the circle of equal altitude is usually several thousand nautical miles (suppose $h = 50°$, then $z = 90° - h$, $z_b = 40° = 2400'$), it is very difficult to plot on a chart to obtain the position with 1 nautical mile accuracy. Therefore, a new method must be found.

Using the Distance Difference Method to Draw the Observation Ship Position

The observation position is always near the estimated position, so as long as a small section of the ship position line near the estimated position is drawn, the ship observation position will be included. Because the radius of the circle of equal altitude is very large, a small arc can be considered a straight line, called the ship position line.

Fig. 4.13 Ship and the
substellar points of the
celestial object

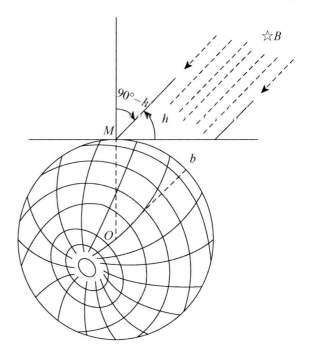

According to these, the deduced ship position is used as a bridge to draw the ship
position line.

As shown in Fig. 4.14, b is the substellar point, c is the ship calculated position,
mm' is the position arc, and cb is the azimuth line. Since the coordinates of two points
c and b are known, the azimuth A_c and distance z_c can be obtained by calculation.

MM' is the tangent line of the ship's position arc mm' at point K, which is perpendicular to azimuth line cb, and MM' is the desired ship's position line. To obtain the

Fig. 4.14 Ship positioning
by distance difference
method

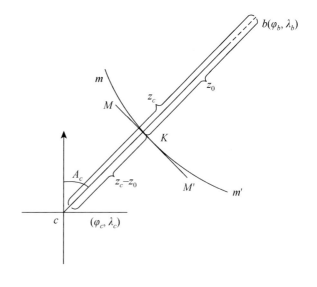

Fig. 4.15 Ac, Zc obtained
by solving spherical triangle

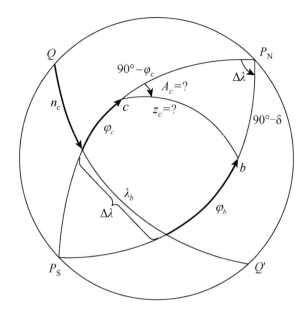

position line, the point K should first be determined. The determination of point K is based on the deduced ship's position c, azimuth A_c and distance difference z_c-z_0, where z_0 is obtained by observation, and A_c and z_c are obtained by solving the resolution of spherical triangles. Figure 4.15 show the Earth, point c is the calculated ship, b is the substellar point, and P_N is the Arctic. The connection of these three points forms a spherical triangular $\triangle cP_N b$, where three conditions are known: $cP_N = 90° - \varphi_c$, $P_N b = 90° - \delta$ (δ is the celestial declination, equal to the latitude of substellar point b), and $\angle cP_N b = \Delta\lambda$(equal to the longitude difference between point b and c). Based on these, A_c and z_c can be solved.

Similarly, the A_c and z_c of another observed celestial body can be calculated. Then, draw two position lines on the chart, the intersection of which is the celestial observation point M (see Fig. 4.16). Flow charts were used to illustrate the process of celestial positioning (Fig. 4.17).

4.4 Satellite Positioning Principle

Satellite navigation and positioning is currently the most influential modern positioning technology, but its history is not long, only more than half a century. Different from celestial navigation, which takes natural celestial bodies as external references, satellite navigation employs satellites as external references. For the principle of positioning, celestial navigation mainly adopts the angular-dependent positioning principle (star atlas matching is not discussed here), while satellite navigation adopts many different positioning methods, such as the distance-dependent positioning principle [9].

Fig. 4.16 Drawing on the
chart to solve the position
resolution

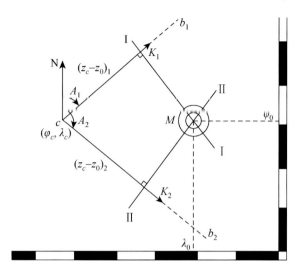

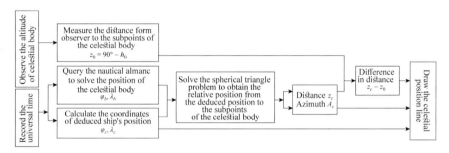

Fig. 4.17 The flow chart of the process of celestial positioning

4.4.1 GNSS Distance-Dependent Positioning Principle

At present, the influential satellite navigation systems in the world include the GPS
of the United States, GLONASS of Russia, Galileo of Europe and Beidou of China.
Each constellation of the GNSS system is designed to consist of 24 or more satellites
to ensure that signals from at least four satellites can be received at any position on
the earth.

Positioning Principle of the Three Spheres Intersect

The direct positioning principles of the four major international satellite navigation
systems, GPS, GLONASS, Galileo and Beidou satellite navigation systems, are basically the same. They all use the geometric principle of three spheres intersecting to

achieve positioning. Its simple process is as follows (Fig. 4.18): the user receiver receives more than three satellite signals at the same time, measures the distances between the receiver and three satellites, and calculates the coordinates of the satellite in the space; then, it takes the satellite as the center of the sphere and the distance from the satellite to the user as the radius, obtaining three user spherical position surfaces; and based on three sphere intersections, it obtains two intersection points. Excluding an unreasonable point, the remaining point is the user's position [10].

On a plane or sphere, the position line at the same distance from the known reference point is a circle; the position surface at the same distance from the known reference point in three-dimensional space is a sphere. The analytic expression of the three-dimensional position function can be expressed as:

$$h_{pk}\left(R^{\gamma}_{\beta\alpha}, R^{\gamma}_{\beta t_k}\right) = d_i$$

$$= \sqrt{\left(x^{\gamma}_{\beta\alpha} - x^{\gamma}_{\beta t_k}\right)^2 + \left(y^{\gamma}_{\beta\alpha} - y^{\gamma}_{\beta t_k}\right)^2 + \left(z^{\gamma}_{\beta\alpha} - z^{\gamma}_{\beta t_k}\right)^2}, k = 1, 2, \cdots, N$$

(4.10)

where $R^{\gamma}_{\beta\alpha} = (x^{\gamma}_{\beta\alpha}, y^{\gamma}_{\beta\alpha}, z^{\gamma}_{\beta\alpha})$ and $R^{\gamma}_{\beta t_k}(x^{\gamma}_{\beta t_k}, y^{\gamma}_{\beta t_k}, z^{\gamma}_{\beta t_k})$ are the position vectors of the position point α and the reference point t_k relative to the coordinate system β and projected to the coordinate system γ.

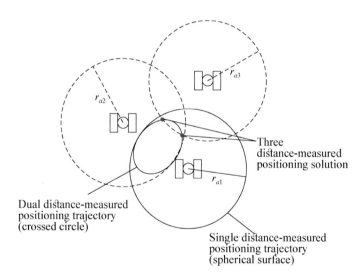

Fig. 4.18 Locational solution trajectory based on different ranging information

Principle of GNSS Pseudorange Positioning

The GNSS pseudorange positioning principle refers to measuring the time delay of signal propagation between the satellite and receiver when the space-borne clock and user clock are synchronized, and the distance between the satellite and user r_{as} is

$$r_{as} = c\tau \qquad (4.11)$$

where c is the speed of light. Considering the clock error between the satellite clock and the user receiver, the measured distance contains errors, so it is called the pseudorange ρ_a^s,

$$\rho_a^s = r_{as} + c\Delta t_u \qquad (4.12)$$

For satellite k, the pseudorange measured by the user receiver is

$$\rho_a^{s_k} = r_{as_k} + c\Delta t_u \qquad (4.13)$$

$$r_{as_k} = \sqrt{\left(x_{ea}^e - x_{es_k}^e\right)^2 + \left(y_{ea}^e - y_{es_k}^e\right)^2 + \left(z_{ea}^e - z_{es_k}^e\right)^2} \qquad (4.14)$$

where $(x_{ea}^e, y_{ea}^e, z_{ea}^e)$ is the position coordinate of the user in the Earth coordinate system, $(x_{es_k}^e, y_{es_k}^e, z_{es_k}^e)$ is the position coordinate of satellite k in the Earth coordinate system, and Δt_u is the clock difference. For the calculation of four unknown quantities $x_{ea}^e, y_{ea}^e, z_{ea}^e, \Delta t_u$ it is necessary to measure the pseudorange of the user to four satellites, and four independent equations can be obtained, namely,

$$\rho_a^{s_k} = \sqrt{\left(x_{ea}^e - x_{es_k}^e\right)^2 + \left(y_{ea}^e - y_{es_k}^e\right)^2 + \left(z_{ea}^e - z_{es_k}^e\right)^2} + c\Delta t_u \quad (k = 1, 2, 3, 4) \qquad (4.15)$$

The position of the satellite can be obtained by solving the ephemeris of the satellite, and the above four unknown quantities can be obtained by solving the simultaneous equation set so that the position of the user can be obtained accurately.

In addition to providing three position coordinates and precise times, the GPS receiver can also provide three velocity components. The rate of change of the pseudorange can be obtained by measuring the Doppler shift of the carrier frequency of the signal wave, and four other equations can be established, namely,

$$\dot{\rho}_a^{s_k} = \frac{\left(x_{ea}^e - x_{es_k}^e\right)\left(\dot{x}_{ea}^e - \dot{x}_{es_k}^e\right) + \left(y_{ea}^e - y_{es_k}^e\right)\left(\dot{y}_{ea}^e - \dot{y}_{es_k}^e\right) + \left(z_{ea}^e - z_{es_k}^e\right)\left(\dot{z}_{ea}^e - \dot{z}_{es_k}^e\right)}{\sqrt{\left(x_{ea}^e - x_{es_k}^e\right)^2 + \left(y_{ea}^e - y_{es_k}^e\right)^2 + \left(z_{ea}^e - z_{es_k}^e\right)^2}}$$

$$= c \cdot \Delta \dot{t}_u \quad (k = 1, 2, 3, 4) \qquad (4.16)$$

where $\dot{x}^e_{es_k}$, $\dot{y}^e_{es_k}$, $\dot{z}^e_{es_k}$, and $x^e_{es_k}$, $y^e_{es_k}$, $z^e_{es_k}$, are the velocity and position coordinates of the satellite in the Earth coordination system, respectively; $\dot{\rho}^{s_k}_a$ are obtained by the measured Doppler shift; $\dot{x}^e_{ea}$, $\dot{y}^e_{ea}$, $\dot{z}^e_{ea}$, is the user's three-dimensional velocity; and $\Delta \dot{t}_u$ is the clock frequency drift. The three-dimensional velocity of the user can be obtained by solving the simultaneous equations.

4.4.2 Principle of Satellite Positioning Based on Range Increment

The earliest satellite navigation system was the Transit Satellite Navigation system of the United States, also known as the Navy Satellite Navigation system (NNSS). The system was put into operation in 1964, using the distance incremental positioning mode of Doppler shift, and can provide global coverage. As GPS matured in the 1990s, the system was stopped.

As shown in Fig. 4.19, the transit satellite navigation system employs the single point positioning method. The system determines the position of the carrier based on the known satellite position (from the satellite ephemeris) and the variation in the frequency of the satellite transmitted signal observed and the precise time. Considering the satellite moving around the Earth as a source of radio waves, point M is the observer on the Earth's surface. According to the Doppler principle, the change in radial distance between the satellite and observer will produce the Doppler effect.

As shown in Fig. 4.20, the projection of the position of satellite s at any time in the Earth coordinate system $r^e_{es} = (x_s, y_s, z_s)$ can be obtained based on the satellite ephemeris, while the projection of the position of observer b in the Earth coordinate

Fig. 4.19 Transit satellite orbit

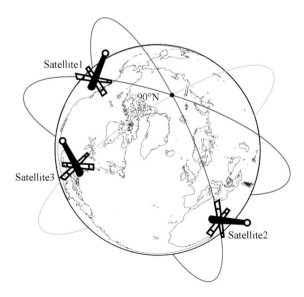

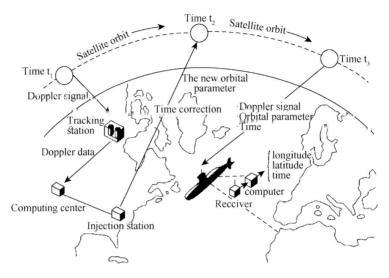

Fig. 4.20 Principle of satellite positioning based on range increment

system is unknown. If the approximate value of the observer's position coordinate is
$r^e_{eb} = (x, y, z)$, the actual coordinates of the carrier can be recorded as:

$$x = x_0 + \Delta x \tag{4.17}$$

$$y = y_0 + \Delta y \tag{4.18}$$

$$z = z_0 + \Delta z \tag{4.19}$$

where Δx, Δy, and Δz is the corrected number of observer coordinates to be sought.
Set the distance between the observer and satellite s at moment i as follows:

$$S_i(x, y, z) = \left[\left(x_{S_i} - x \right)^2 + \left(y_{S_i} - y \right)^2 + \left(z_{S_i} - z \right)^2 \right]^{\frac{1}{2}} \tag{4.20}$$

The first-order Taylor expansion of the upper form at (x_0, y_0, z_0) is obtained as
follows:

$$S_i(x, y, z) \approx S_i(x_0, y_0, z_0)$$
$$+ \left. \frac{\partial S_i(x, y, z)}{\partial x} \right|_{\substack{x = x_0 \\ y = y_0 \\ z = z_0}} \Delta x + \left. \frac{\partial S_i(x, y, z)}{\partial y} \right|_{\substack{x = x_0 \\ y = y_0 \\ z = z_0}} \Delta y + \left. \frac{\partial S_i(x, y, z)}{\partial z} \right|_{\substack{x = x_0 \\ y = y_0 \\ z = z_0}} \Delta z \tag{4.21}$$

Then, the distance difference between two consecutive moments is:

$$\Delta S_{i,i+1}(x, y, z) = S_{i+1}(x, y, z) - S_i(x, y, z)$$
$$= S_{i+1}(x_0, y_0, z_0) - S_i(x_0, y_0, z_0) + A\Delta x + B\Delta y + C\Delta z$$

$$(4.22)$$

In the formula,

$$A = \frac{x_0 - x_{S_{i+1}}}{S_{i+1}^\circ} - \frac{x_0 - x_{S_i}}{S_i^\circ}$$

$$B = \frac{y_0 - y_{S_{i+1}}}{S_{i+1}^\circ} - \frac{y_0 - y_{S_i}}{S_i^\circ} \qquad (4.23)$$

$$C = \frac{z_0 - z_{S_{i+1}}}{S_{i+1}^\circ} - \frac{z_0 - z_{S_i}}{S_i^\circ}$$

According to the Doppler theorem, the relative velocity of the observer and satellite is:

$$v_{bs} = \frac{c}{f_t}\Delta f_{tr} \qquad (4.24)$$

where c is the velocity of the radio wave, f_t is the frequency of the radio wave, and Δf_{tr} is the Doppler shift between the transmitter and receiver.

Therefore, the relative displacement between the observer and the satellite is:

$$\Delta \tilde{S}_{i,i+1}(x, y, z) = v_{bs} \cdot (t_{i+1} - t_i) = \frac{c}{f_t}\Delta f_{tr} \cdot \left[N - (t_{i+1} - t_i)\right] \qquad (4.25)$$

In the formula, N is integer ambiguity.

From (4.17) to (4.20):

$$A\Delta x + B\Delta y + C\Delta z = L \qquad (4.26)$$

In the formula,

$$L = S_{i+1}(x_0, y_0, z_0) - S_i(x_0, y_0, z_0) - \frac{c}{f_t} \cdot \Delta f_{tr}\left[N - (t_{i+1} - t_i)\right] \qquad (4.27)$$

The above formula (4.26) is the error equation for satellite Doppler observations. A, B, C and L can be calculated from observations and known data. The equation contains only 3 undetermined unknowns, Δx, Δy, and Δz. It takes approximately 18 min to observe a satellite passing through, that is, it rises from the ground level and falls from the ground level through the highest point. Therefore, taking 30 s as the integral interval, 20 or 30 error equations can be obtained. According to the least squares method, we can determine a set of observer coordinate corrections with the least sum of residual squares. When several satellite passes are observed, the joint solution of all passes can also be obtained according to the sequential adjustment method. After the coordinate correction is obtained, the estimated position of the observer is obtained according to formulas (4.17), (4.18), and (4.19).

These are the basic principles of satellite positioning based on range increments [11].

4.4.3 Main Global Satellite Navigation System

GNSS has become the focus of current navigation technology competition among different countries. GNSS has a huge impact on civilian applications, from personnel applications such as health bracelets, mobile phones, intelligent vehicles, and shared bikes to national applications such as earth ocean surveys, crust movement, meteorological information measurements, and urban traffic management. Satellite navigation technology has widely and profoundly affected human social life.

Global Positioning System (GPS)

GPS is the second-generation US satellite navigation and positioning system. The first experimental satellite was launched in 1978. In December 1993, GPS had an initial operational capability. The system achieved Full Operational Capability (FOC) in July 1995. The system is the third huge space plan in the US after the Apollo moon landing program and the space shuttle program, which had taken nearly 20 years and cost more than 10 billion US dollars to build. GPS has become the infrastructure of the US National Navigation Information Service. In Gulf War 1991, GPS profoundly influenced the development of warfare and weapons in modern warfare.

Early GPS adopted P (Y) and C/A code policies to artificially reduce the C/A code accuracy. In 1996, the GPS ground segment began to implement precision improvement and a wide-area GPS improvement plan to continuously improve GPS accuracy and replace failed satellites with new satellites (BlockIIR). In May 2000, US President Bill Clinton announced the termination of the selection availability (SA) policy, which improved the GPS civilian accuracy, with a positioning accuracy better than 20 m.

In 2000, a GPS modernization program was implemented, replacing new navigation satellites BlockIIRM and BlockIIF and developing SAASM and GPSIII systems. GPSIII has made significant adjustments to all GPS segments to significantly improve military and civilian service performance. In December 2004, in response to the European development of the Galileo system and international anti-terrorism, the US issued the "New American Star-based Positioning, navigation and timing Policy". In addition, facing the challenge of a navigation war, GPS modernization was implemented. GPS III's first satellite was successfully launched in December 2018. Compared with the second-generation GPS satellite, the satellite has extended to 15 years, life was doubled, accuracy was tripled, and anti-interference ability was increased by 8 times.

The navigation signals can also be quickly turned off to specific geographical locations according to the actual needs. In 2026, the GPS system plans to launch its first GPS IIIF satellite, with 22 satellites scheduled to be built, and the launch is completed in 2033.

GPS is widely used in aircraft, automobiles and ships and is widely used in aircraft route guidance and approach landing, ship marine navigation, automobile autonomous navigation and tracking positioning, urban intelligent transportation management and other fields. In addition, it also plays a huge role in the emergency rescue of police, hospital and fire protection, target tracking and emergency dispatch. At present, GPS has become the most influential and extensive positioning system worldwide with its global, high precision, automation, high efficiency and all-weather positioning service functions.

GLONASS System

The GLONASS system is the second generation of the Soviet satellite navigation and positioning system, developed in the 1970s, and the first satellite was launched in 1984. In January 1996, it was put into overall operation. The GLONASS system adopts the FDMA system, and its anti-interference ability is better than that of the GPS system, but its single-point positioning accuracy is less than that of the GPS system. After the collapse of the Soviet Union, due to various factors, the Russian space industry experienced a series of setbacks, and the number of satellites in orbit continuously decreased. In 2002, only 18% of the space network was available, and the system positioning accuracy decreased to 35 m. By 2007, only 6 satellites were in orbit. Over the past 20 years, with Russia's national strength constantly recovering, the system has recovered rapidly and achieved great development. In July 2011, Russia fully resumed its global operational services. In 2012, there were 29 satellites in orbit, with a positioning accuracy of 2.8 m. There are currently 31 GLONASS navigation satellites in orbit. In the future, CDMA signals will be introduced to enhance system performance. By 2020, the system navigation accuracy reaches 0.6 m. GLONASS satellites in orbit will increase to above 36 by 2030.

Galileo System

Due to the important strategic value and huge commercial interests of GNSS, to eliminate the serious dependence on the GPS system, the European Commission (EC) proposed the establishment of a European independent global satellite navigation and positioning system. In March 2003, the Galileo project was approved by the European Union. The Galileo system uses 27 working satellites, two ground stations and three backup satellites with better design accuracy than GPS.

In April 2010, after years of consultation and debate, the European Parliament passed the systematic final scheme, marking the official launch of the Galileo project for infrastructure construction. In response to the competition in China's Beidou system, Europe has accelerated its system construction. In March 2013, the space and ground infrastructure worked together and completed the first positioning of ground users. The average horizontal positioning accuracy was 8 m, the localization efficiency was 95%, and the timing accuracy was 10^{-9} s. By the beginning of 2020, 22 of the 26 satellites in orbit worked functionally and provided service. All satellites are expected to enter orbit by the end of 2020.

In December 2016, the EC officially announced the launch of the system. Initial services include open services, licensing services, and search and rescue services. Open Services (OS) is for the mass market, fully interoperable with GPS, to provide users with more accurate and reliable services. The Public Regulated Services (PRS) are encrypted, more robust services, making services available to government-authorized users such as civil defense, fire protection, and police departments. Search and Rescue Services (SAR) means the contribution of international search and rescue services. Using the Galileo system and other GNSS-based SAR services, the user located the distress beacon when emergencies at sea or in the wilderness for only 10 min with position accuracy within 5 km. With the continuous complement of satellites and the improvement of ground facilities, its performance will continue to improve.

Beidou System

In 1983, Chen Fangyun, a famous Chinese space expert, first proposed the idea of using two geostationary orbit communication satellites to achieve rapid regional navigation and positioning in China. In 1989, a binary star positioning demonstration test based on a communication satellite proved the correctness and feasibility of the technical system.

In the late twentieth century, China began to explore the development of satellite navigation systems suitable for national conditions and gradually formed the three-step development strategy of the Beidou Navigation Satellite System (BDS) [12]:

(1) The first step is the construction of the Beidou-I (BD I) system. Construction started in 1994; in December 2000, two satellites were launched (at 140°E and 80°E). In 2003, the system was put into operation, forming an active service capability covering China. The uncalibration accuracy is 100 m, and the calibration accuracy is 20 m.

(2) The second step is to launch the construction of the Beidou-II (BD II) satellite navigation system in 2004. Completed in 2012. The regional communication service capability of navigation, timing and short message covering the Asia–Pacific region is officially realized, with a positioning accuracy of 10 m, a timing accuracy of 50 ns and a speed measurement accuracy of 0.2 m/s.

(3) The third step is to launch the construction of the Beidou-III global navigation satellite system in 2012, which was completed in July 2020. The 35 satellites

Table 4.1 Characteristics of the Earth orbit in GNSS

Constellation	Number of track planes	Radius/ km	Altitude/ km	Period	Number of orbital circles per solar day	The substar trajectories repeat period/solar day	Inclination of orbit
GPS	6	26 580	20 180	11 h 58 min	2	1	55°
GLONASS	3	25 500	19 100	11 h 15 min	2.125	8	64.8°
Galileo	3	29 620	23 220	14 h 5 min	1.7	10	56°
BDII	3	27 840	21 440	12 h 52 min	1.857	7	55°

include 5 synchronous orbit satellites, 3 inclined synchronous orbit satellites and 27 medium-circular orbit satellites, becoming a high-precision, all-time satellite navigation and positioning system and providing services to the world.

Compared with other satellite navigation systems, the main features of the Beidou system are as follows: ① The space section of theBeidou system adopts a mixed constellation composed of three orbital satellites, with more high-orbit satellites and strong occlusion resistance, especially at low latitudes. ② The Beidou system provides navigation signals at multiple frequency points, which can improve the service accuracy through the combination of multifrequency signals. ③ The Beidou system innovatively integrates navigation and communication capabilities, with multiple capabilities and other service capabilities, such as positioning and navigation timing, star-based enhancement, ground-based enhancement, precision single-point positioning, short message communication and international search and rescue.

The orbital characteristics of each satellite navigation system are shown in Table 4.1.

4.5 Radio Positioning Principle

Radio navigation uses radio technology to locate and guide the carrier along a predetermined route. In essence, satellite navigation is also a kind of radio navigation. Radio navigation has a relatively longer history, dating back to the early twentieth century. From the initial radio direction finder, the Omega global radio navigation system (disabled in 1997), various types have been developed, such as aviation distance measure equipment (DME) beacons and VHF omnidirectional

Fig. 4.21 Hyperboloid
position line

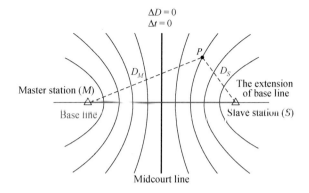

radio beacons (VOR). The long range navigation system (LORAN). to civilian-based mobile phone signals, wireless local area networks (WLAN), Wi Fi, wireless personal local area networks (WPAN), RFID, and ultrawide band (UWB) in recent years. Radio positioning technology for communication, television signal and radio broadcast signal. This section provides a brief principle of radio positioning based on the direct positioning principle.

4.5.1 Hyperbolic Positioning Based on Ranging Difference

Measuring the Distance Difference Between the Observer and the Two References

From the analytic geometry, if the observer and the two reference points are in the same plane (or sphere), the distance difference between the observer and the two known reference points can be measured, and the position line of the observer is two hyperbolic curves. The two reference points are the focus of the hyperbolic curve (Fig. 4.21) [1, 5]. If the observer and two reference points are considered in three-dimensional space, the position surface of the observer is two hyperboloids. Similarly, if it is possible to measure the sum of the distance between the observer and the two reference points, as seen from the analytic geometry, the position line of the observer is elliptical, and the two reference points are the focus of the ellipse (Fig. 4.22). If the observer and the two reference points are considered in three-dimensional space, the position surface of the observer is an ellipsoid.

Fig. 4.22 Ellipsoid position
line

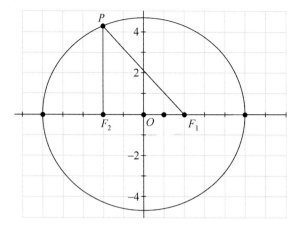

If the distance difference between the observer $\mathbf{r}_{\beta\alpha}^{\gamma}$ and the two known reference
points $\mathbf{r}_{\beta t_1}^{\gamma}, \mathbf{r}_{\beta t_2}^{\gamma}$ is Δd, the analytic expression of the observer's position function can
be expressed as:

$$
h_{pi}\left(\mathbf{r}_{\beta\alpha}^{\gamma}, \mathbf{r}_{\beta t_1}^{\gamma}, \mathbf{r}_{\beta t_2}^{\gamma}\right) = \Delta d_i = \sqrt{\left(x_{\beta\alpha}^{\gamma} - x_{\beta t_1}^{\gamma}\right)^2 + \left(y_{\beta\alpha}^{\gamma} - y_{\beta t_1}^{\gamma}\right)^2 + \left(z_{\beta\alpha}^{\gamma} - z_{\beta t_1}^{\gamma}\right)^2}
$$
$$
-\sqrt{\left(x_{\beta\alpha}^{\gamma} - x_{\beta t_2}^{\gamma}\right)^2 + \left(y_{\beta\alpha}^{\gamma} - y_{\beta t_2}^{\gamma}\right)^2 + \left(z_{\beta\alpha}^{\gamma} - z_{\beta t_2}^{\gamma}\right)^2}
$$

(4.28)

Loran C System

The hyperbolic positioning system is also known as the range difference system. The
representative systems are the Loran A system, Deca system, Loran C system and
so on [13].

Loran C system is a longwave pulse phase hyperbolic radio navigation and positioning system. Applied in the 1960s, the center frequency of the carrier wave is 100
kHz. The system range based on the ground wave is 900–1300 nautical miles, and the
positioning accuracy is 0.4.1.2 nautical miles. The system has the advantage of all-
weather continuous real-time navigation and positioning. Loran C system currently
covers most of the Northern Hemisphere, and its distribution is shown in Fig. 4.23.

The Loran C receiver is used to determine the time difference or phase difference of
the radio waves transmitted from master station M and slave station S to the receiving
point. According to the propagation speed c and wavelength λ of radio waves, the time
difference $\triangle t$ and phase difference φ can be converted into the distance difference
between the observer and the two navigation stations. The observer position line
determined by the distance difference is a hyperbolic curve focusing on the M and S of

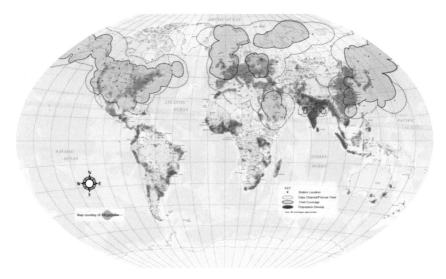

Fig. 4.23 Loran C system chain distribution

two navigation stations. If two distance differences can be measured, two hyperbolic position lines can be obtained, and one of the intersecting points is the position of the observer (Fig. 4.24).

The Loran C receiver is mainly composed of an antenna, antenna coupler and receiving indicator. The antenna and the antenna coupler are attached together to the higher place outside of the cabin. The receiving indicator is connected to the coupler by a dedicated cable. The signal received by the antenna is filtered, amplified and impedance converted to the receiving indicator. The receiving indicator is installed in the cabin, which consists of the receiving channel, timer, data processing, keyboard and display parts.

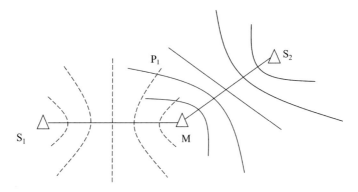

Fig. 4.24 Positioning by ranging difference method

4.5.2   Single Azimuth-Distance-Dependent Positioning

Polar Coordinate Positioning

Single azimuth-distance-dependent positioning, also known as the polar coordinate positioning method, obtains the position of the observer based on the intersection of the azimuth position line and the distance position line of the reference point at the same time by simultaneously observing the azimuth and distance of a single external reference. Single reference azimuth-distance-dependent positioning is widely used, such as using radar to observe the azimuth and distance of the reference to achieve positioning.

Tactical Air Navigation System (TACAN)

The TACAN system is based on the principle of single point positioning using a polar coordinate system. The TACAN system uses the inquiry response mode to carry on the ranging. The airborne TACAN equipment randomly transmits the inquiry pulse pair after the ground TACAN equipment receives and then sends out the reply pulse in the form of the pulse pair. Based on the time it takes to send an inquiry pulse to receive an answer pulse and the speed of radio waves, the airborne TACAN equipment can calculate the distance from the aircraft to the ground station. The angle measurement of the TACAN system is achieved by measuring the phase relationship between the reference pulse signal and the pulse envelope signal. When the aircraft is in different azimuths of the TACAN ground station, there is a different phase relationship between the reference pulse signal and the pulse envelope signal received by its airborne TACAN equipment, so that the azimuth angle of the aircraft relative to the TACAN ground station can be determined.

The TACAN system is composed of ground TACAN equipment and airborne TACAN equipment. Airborne equipment includes radio transceivers, antennas, controls and display devices. Ground includes radio transceivers, antennas, monitors and control devices. The system works in the UHF frequency band between 962 and 1213 MHz, with frequency intervals of 1 MHz and a total of 252 channels. Takon works by selecting the channel. Each channel corresponds to two (transmitter and receiver) carrier frequencies, with two frequencies separated by 63 MHz. The signal propagation mode is mainly distance-of-sight transmission. A typical TACAN system works at a distance of 370 km, ranging accuracy of approximately ± 200 m and angular accuracy up to ± 1°.

The greatest advantage of single azimuth-distance-dependent positioning is that the intersection θ of the position line is always equal to 90°; only the azimuth and distance of a reference can be measured in the observation range to determine the carrier position. The positioning method is simple and fast, and no running fix is needed.

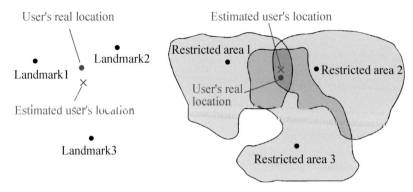

(a) Neighborhood averaged position result   (b) The restricted area intersection

Fig. 4.25 Proximity positioning using multiland marks

4.5.3 Civilian Radio Navigation System

Positioning System Based on Adjacent Positioning

Proximity location is the simplest direct location and is mainly used in short-distance radio positioning. That is, if the user receiver receives a radio signal from the external reference (transmitter), the external reference position can be regarded as the user position. Using multiple external references can improve the accuracy of proximity positioning. The simplest approach is to set the positioning solution to the average position of the external reference point. The closer the user is to the reference point transmitter, the more accurate the proximity positioning method will be.

The more advanced method of proximity positioning is to adopt restricted area intersection positioning. The area covered by each transmitter can determine a limit area. The restricted area is not simply circular due to topography and obstacles. For example, when a landmark is observed, the user location is limited to the restricted area of that landmark. By observing multiple landmarks, it can be determined that the user location is within the intersection area of these landmark restricted areas, and the center of the intersection area can be used as the location point (Fig. 4.25).

Proximity localization can be applied in a variety of applications, such as bluetooth low energy (BLE), radio frequency identification (RFID), very short range radio signals, wireless personnel area networks (WPAN), wireless local area networks (WLAN), etc. The positioning accuracy can reach several to tens of meters.

AGPS Navigation Technology

Assisted GPS (AGPS) technology is a "GPS + base station" positioning technology that uses the base stations of communication networks with GPS satellites to make mobile phone positioning faster. There are two positioning methods adopted by the mobile phone: one is based on GPS positioning, and the other is based on the mobile operation network base station. The GPS mode uses the GPS positioning module of mobile phones to realize mobile positioning, which usually works well outdoors. However, it is less effective where the satellite signals are covered, e.m. in the indoor environment. The base station mode can obtain a more accurate position based on the measured range from the location of the base station to the mobile phone. The latter does not require the mobile phone to have GPS positioning capability, but the accuracy largely depends on the distribution of the base station and the coverage of the signal.

Wi-Fi Positioning Technology

The commonly used WiFi positioning in the civilian field uses the Wi-Fi signal for range positioning. Similar to the positioning mode of mobile phone base stations, the positioning principle of wireless network Wi-Fi requires collecting the location information of Wi-Fi hotspots, which is generally relatively fixed. Once a Wi-Fi hotspot is powered on, it emits a signal to its surroundings, and the signal contains the unique global ID (MAC) of the hotspot. Even if the user cannot establish a connection with hot spots successfully, he can still receive the signal, detect its multitude, and send the information to the network server. According to the information, the server can query the database record to obtain the hotspot coordinates. Then, either based on multipoint positioning or adjacent positioning, the specific location of the client can be obtained, and then the coordinates will be sent to the client through the network. The more Wi Fi hotspot signals the user receives, the more accurate the positioning will be.

4.6 Acoustic Positioning System

Hydroacoustic baseline localization is a technical means of achieving maritime localization by determining the propagation time or phase difference of acoustic wave signals. The hydroacoustic baseline positioning system, hereinafter referred to as the acoustic positioning system (APS), mainly refers to the underwater sound system that can be used for accurate positioning and navigation of local areas. t forms an acoustic matrix by placing multiple acoustic receivers or transponders in the sea area.

The long baseline system (LBL) and the short baseline system (SBL) obtain the distance from the substrate by measuring the acoustic wave propagation time to calculate the target position. The ultrashort baseline positioning system (USBL) realizes the positioning solution through phase measurement. The above system is characterized by installing beacons, transponders, responders, or multiple arrays of transponders that can transmit acoustic signal signs on the ocean floor as underwater reference points. This kind of acoustic navigation and positioning system has many similarities with the radio navigation system with the shore radio beacon as the reference point, that is, by measuring the distance between the reference point and the carrier to determine the distance between the two and then solving the position of the carrier to realize navigation and positioning. However, the acoustic positioning system is vulnerable to underwater noise interference, and excessive seawater bubbles will seriously attenuate the signal. Therefore, it cannot be relied on to provide continuous positioning, and it is generally used as part of an integrated navigation system [14].

4.6.1 Long Baseline Acoustic Positioning System

As shown in Fig. 4.26, the LBL includes two parts: a transceiver transducer mounted on a ship or underwater carrier and a series of known transponders fixed to the seabed, and distances between them form a baseline.

The long baseline lengths are usually between a few hundred meters and several kilometers. With the development of technology, the baseline length is still further expanding. The system is called a long baseline system relative to an ultrashort and short baseline system. The LBL system measures the distance between the transceiver and the transponder using the front or rear intersection of the measured target positioning, so the system is irrelevant to the depth, and the carrier does not have to install attitude measuring equipment and gyrocompass.

Figure 4.26 show the navigation and positioning principle of surface carriers using three transponders. Suppose the z-coordinate of the surface carrier transducer is 0, and T_1, T_2 and T_3 are hydroacoustic transponders with known positions, which are

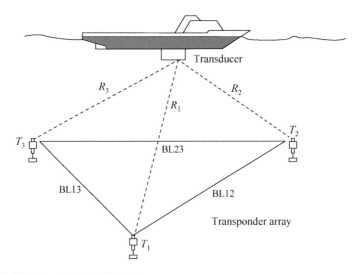

Fig. 4.26 Working principle of LBL Systems

(x_1, y_1, z_1), (x_2, y_2, z_2), and (x_3, y_3, z_3), respectively, and BL12, BL13 and BL23 are the baselines. When the carrier transducer emits a "ping" digital modulated audio pulse at a frequency of up to 40 kHz. The LBL transponder receives a user "ping" signal, and after a fixed interval, responds with a similar "ping" signal. The carrier transducer measures the time difference of the received signal and the corresponding processing and can obtain the distance R_i between the carrier and transponder i, namely, R_1, R_2 and R_3. The position (x, y) of the surface vehicle can be solved by three R_i cubic equations.

$$
\begin{aligned}
x &= \frac{D_{12}(y_3 - y_2) + D_{23}(y_1 - y_2)}{x_1(y_3 - y_2) + x_2(y_1 - y_3) + x_3(y_2 - y_1)} \\
y &= \frac{D_{12}(x_3 - x_2) + D_{23}(x_1 - x_2)}{x_1(y_2 - y_3) + x_2(y_3 - y_1) + x_3(y_1 - y_2)}
\end{aligned}
\tag{4.29}
$$

where $D_{ij} = \frac{1}{2}\left(x_i^2 - x_j^2 + y_i^2 - y_j^2 + z_i^2 - z_j^2 + R_j^2 - R_i^2\right)$

Obviously, for navigation and positioning with three or more transponders, the accuracy depends on the ranging accuracy and requires compensation for the carrier motion during the ranging process. For underwater unmanned vehicles (UUVs) with unknown depths, four transponders are generally used.

If the operating frequency of the hydroacoustic system is 10–15 kHz, the maximum operating distance of the transponder is 10 km, and considering that the sound wave travels at 1500 m/s, the time interval between sending and receiving the "ping" signal will exceed 10 s. During this period, if the carrier moves tremendously, the carrier movement must be compensated during positioning. Both underwater vehicles and ships are generally equipped with a dead reckoning system or an inertial navigation system, so the short-term position changes can be determined relatively accurately.

The advantages of an LBL system are as follows: ① the measurement is independent of the water depth value; and ② a very high relative positioning accuracy can be obtained due to many redundant observations. ③ The transducer of the LBL system is small and easy to install and remove in practice. ④ It can be used for both surface and underwater vessel positioning.

The disadvantages are as follows: ① The overall system is complex, and the operation is cumbersome; ② equipment is expensive; ③ it takes a long time to layout and recycle a huge number of sound base arrays, requiring careful calibration. ④ The determination of the absolute position of the sound base arrays must be obtained by ship ranging at several known locations.

4.6.2 Short Baseline Acoustic Positioning System

Different from the LBL system, the SBL system can not only adopt the "distance-distance" method but also adopt the "azimuth-distance" method and "azimuth-azimuth" method for positioning. The transponder array of the SBL system is not located on the seabed but on the bottom of the carrier. The underwater part of the SBL system requires only one underwater acoustic transponder, which can be installed on the seabed at a known location that serves as an external positioning reference or on a UUV of unknown location that serves as the underwater target for dynamic positioning. The onboard part of the SBL system is a transducer base array installed at the bottom of the ship. The distance between the transducers is generally more than 10 m, which should be accurately measured during the installation to establish an accurate acoustic array coordinate system. The distance and angle relationship between the array coordinate system and the body coordinate system are determined by the conventional measurement method.

When establishing the seabed geodetic reference, according to the relation of the acoustic arrays and the ship, combined with the external sensor observations, such as the center transducer position by GNSS, the ship attitude and heading measured by the motion reference unit (MRU) and the gyrocompass, respectively, the geodetic coordinates of the seabed reference can be calculated.

Figure 4.27 illustrates the configuration of an SBL system. H_1, H_2, and H_3 are the receiving transducers, O is the only transmitting transducer (also the center of the body coordinate system), and the transducers are orthogonally installed. The baseline length between H_1 and H_2 is b_x, pointing to the bow, the x-axis; the baseline length between H_2 and H_3 is b_y, parallel to the y-axis, pointing to the starboard, and the z-axis pointing to the seabed. The SBL system transmits acoustic signals from the transmitting transducer and is received by other transducers. Multiple slantranges are calculated by measuring the propagation time of the different received acoustic paths. The angles between the transmitting acoustic line and the three axes are θ_{m_x}, θ_{m_y} and θ_{m_z} (Fig. 4.28), and Δt_1 and Δt_2 are the time differences between the acoustic signals received by H_1, H_2 and H_2, H_3, respectively.

The "azimuth-distance" method of the SBL is briefly described here. To simplify the problem, the multiple-valued period number of the phase difference measurement is not considered. From Fig. 4.28, we obtain:

$$\left.\begin{array}{l} \cos\theta_{mx} = \frac{c\cdot\Delta t_1}{b_x} = \frac{\lambda\Delta\phi_x}{2\pi b_x} \\ \cos\theta_{my} = \frac{c\cdot\Delta t_2}{b_y} = \frac{\lambda\Delta\phi_y}{2\pi b_y} \\ \cos\theta_{mz} = (1 - \cos^2\theta_{mx} - \cos^2\theta_{my})^{\frac{1}{2}} \end{array}\right\} \quad (4.30)$$

In the formula, $\Delta\phi_x$ and $\Delta\phi_y$ are the phase differences between the signals received by H_1, H_2 and H_2, H_3, respectively.

According to the angle of the spatial line OP and each coordinate system and the length of the OP, the coordinates (x, y, z) of point P can be directly obtained from Fig. 4.28 in the body coordinate system, i.e.,

Fig. 4.27 The configuration of SBL

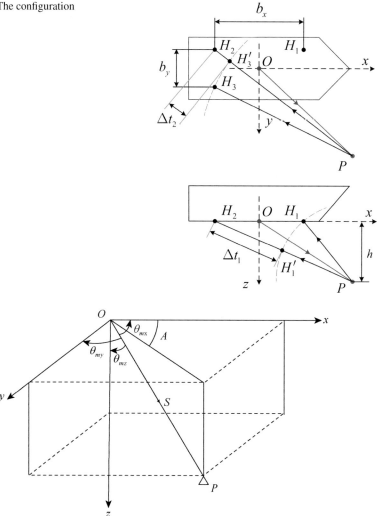

Fig. 4.28 SBL positioning

$$\begin{cases} x = S \cos \theta_{m_x} \\ y = S \cos \theta_{m_y} \\ z = S \cos \theta_{m_z} \end{cases} \tag{4.31}$$

The advantages of SBL are as follows: ① the integrated system is inexpensive; ② the system is easy to operate; and ③ the transducer is small and easy to install.

The disadvantages of the SBL are as follows: ① To achieve high accuracy in deep water, the baseline length generally needs to be greater than 40 m. ② When the system is installed, the transducer must be strictly calibrated on the dock. ③ The short

Fig. 4.29 Schematic
diagram of the USBL array

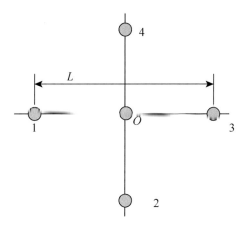

baseline architecture is suitable for underwater vehicles and divers relatively close to the ship. When the distance between the user and the ship exceeds the baseline length, the signal geometry deteriorates, resulting in a decrease in the positioning accuracy.

4.6.3   Ultrashort Baseline Acoustic Positioning System

The acoustic arrays of the USBL system are mounted in a very small housing on the bottom of the ship (Fig. 4.29). The mutual position between the sound units is accurately determined to form the acoustic array coordinate system. Figure 4.28 shows the relationship between the acoustic array coordinate system and the body coordinate system, including the origin position deviation between the two coordinate systems and the installation deviation angle of the three axial directions of the acoustic arrays [15].

The USBL systems can be used for surface ships to determine the relative position of the targets in the water, such as UUV, minesweeper, and frogman.

The USBL system usually operates at a frequency of 10–15 kHz. Unlike the SBL system, the acoustic array of the USBL system is separated at a short distance between units b, only a few centimeters, less than half a wavelength. Taking the phase difference to measure the target orientation will not result in the multiple-valued period number problem. According to Eqs. (4.30) and (4.31), the relative target coordinates are

$$\begin{cases} x = S \cdot \cos\theta_{m_x} = \frac{1}{2}ct\dfrac{\lambda\Delta\phi_x}{2\pi b} = \dfrac{ct\lambda\Delta\phi_x}{4\pi b} \\[4mm] y = S \cdot \cos\theta_{m_y} = \frac{1}{2}ct\dfrac{\lambda\Delta\phi_y}{2\pi b} = \dfrac{ct\lambda\Delta\phi_y}{4\pi b} \end{cases} \quad (4.32)$$

According to formula (4.32), the horizontal bearing angle A between the acoustic line and the bow can be further found at

$$\tan A = \frac{y}{x} = \frac{\Delta\phi_y}{\Delta\phi_x} \quad (4.33)$$

An USBL system can also be used to determine the absolute position of underwater carriers based on seabed-mounted transponders. Based on the known absolute coordinates (x_p, y_p) of the seabed-mounted transponder, the ship heading K (provided by the ship's gyrocompass), and Eq. (4.32), the relative distance D of the transponder to the acoustic array center can be calculated. The absolute position of the carrier can be determined according to the following formula:

$$\begin{aligned} x &= x_p - D\cos(A + K) \\ y &= y_p - D\sin(A + K) \end{aligned} \quad (4.34)$$

Both the SBL and USBL systems need only one underwater transponder (Fig. 4.30), which is much easier to deploy than the LBL system, which needs to deploy a transponder array. The system error is relatively low, but the operation range is short.

The advantages of the USBL are as follows: ① The bottom hydrophone array is less affected by the dynamics of the ship; ② Only one integrated unit is needed, which is easy to install, and it is available for both large and small boats. ③ The integrated system is inexpensive and easy to operate.

The disadvantages are as follows: ① The calibration after system installation needs to be very accurate, and error calibration is difficult; ② The absolute position accuracy of the measurement target depends on the measurement accuracy of the ship heading, attitude and depth provided by peripheral devices.

4.7 Angle Measurement and Range Measurement

Radio navigation, acoustic navigation and optical navigation are navigation technologies that mainly realize positioning by measuring the distance or orientation or the spatial geometric relationship between the carrier and the external reference points. In terms of commonness, the above methods can be classified as wave-based methods,

but the characteristics of the three types of waves are different. The general equation
for the waves contains four basic parameters: amplitude, frequency, time, and phase.
During propagation, the amplitude, frequency, time interval and phase of the wave
may change in relation to navigation parameters such as distance and orientation.
If the angle, distance and speed information can be extracted from this information,
navigation and positioning can be realized. Light has wave–particle duality. The
particle nature of light is mainly used in navigation measurements based on external
references. However, when using an optical gyro to realize inertial navigation, light
fluctuation is utilized.

Angle measurement and ranging are the technical basis of direct positioning, and
the technical methods are diverse and quite different from each other. This section
describes this issue.

4.7.1 Angle Measurement Method

Direction Finding Based on Optical Aiming Mechanical Rotation

Direction finding based on mechanical rotation has two methods: manual reading and
digital output of the photoelectric coder. Sextant (Fig. 4.31) and azimuth instruments
are the most basic angle measurement tools for manual reading. The fan-shaped
Sextant is a precision goniometer composed of three parts: frame, optical system,
angle measuring and reading device. The optical system includes a small telescope,
a translucent and semireflective fixed plane mirror (the horizontal mirror), and a

Fig. 4.31 Sextant,
photograph by Bian H.W.

moving mirror connected to the indicator (the indicator mirror). The scale arc of the sextant is 1/6 of the circumference. When used, the observer holds a sextant and rotates the indicator mirror so that the celestial object and the sea horizon simultaneously appear in the field of view and coincide with each other (Fig. 4.32). The height angle of the object can be read according to the angle of the indicator mirror, whose error is approximately ±1° to ±2°. The sextant can be used to measure the horizontal angle between two targets or to measure the height angle of the sun or other celestial objects to the sea horizon.

Mechanical Rotation and Direction Measurement Based on the Signal

The directionality of the antenna is used to determine the radio wave direction. Rotate the antenna and measure the induction signal strength. According to the signal amplitude, e.g., minimum value, maximum value or comparative value, the signal direction is determined based on the mechanical rotation angle.

(1) **direction determined by the minimum signal**

This method determines the radio wave direction by determining the minimum signal level based on the directionality of the annular receiving antenna (Fig. 4.33). When the radio wave is perpendicular to the annular antenna plane, the received signal is

Fig. 4.32 Sextant principle

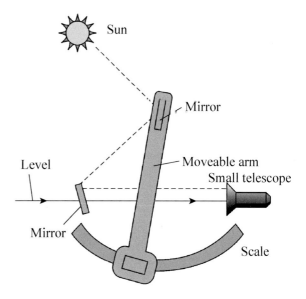

Fig. 4.33 Minimum signal direction

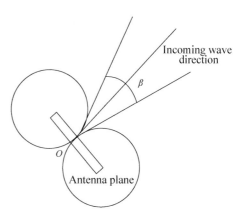

minimal and has a certain angle range (silent angle or dumb point), within which no signal change can be identified.

(2) **direction determined by the maximum signal**

This method determines the radio wave direction by determining the maximum signal level based on the antenna radiating pattern. The narrower the radiating pattern is, the more obvious the signal changes at the maximum value. If the signal near the maximum changes slowly, the direction-finding accuracy is low (Fig. 4.34).

Fig. 4.34 Maximum signal direction

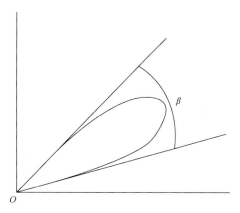

(1) direction determined by Comparative signal

This method takes two antennas with the same radiating pattern simultaneously and determines the direction of the radio wave by comparing the receiving signals. There are two main methods: the amplitude comparison method and the equal strength signal comparison method.

The amplitude comparison method takes two vertical antennas of the same radiating pattern (Fig. 4.35); then,

$$E_1 = E \cdot \cos \theta \tag{4.35}$$

$$E_2 = E \cdot \sin \theta \tag{4.36}$$

where θ is the angle between the incoming wave and an antenna plane, and the angle θ depends on $\tan\theta = E_2/E_1$.

The equal strength signal comparison method adopts two overlapping antennas of the same radiating pattern (Fig. 4.36). When the two antennas are aligned with the incoming wave direction, the strength of the two received signals is equal; otherwise, the signal strength is different, which can be used to determine the incoming wave direction.

Direction Finding Based on the Sensor Arrays

This method realizes signal direction finding by controlling and measuring the phase of the sensor array signal, measuring the signal without the mechanical rotation of the sensor array, which is an important direction-finding method in modern fields of communication and detection systems. The method is applied to the SBL, USBL acoustic positioning system and multiantenna satellite attitude detection system [16].

Fig. 4.35 The amplitude
comparison method

Antenna plane

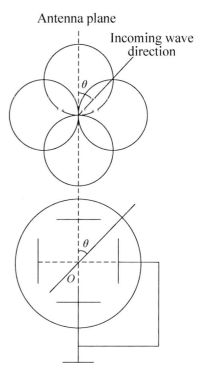

Incoming wave
direction

Fig. 4.36 The equal strength
signal comparison method

Incoming wave direction

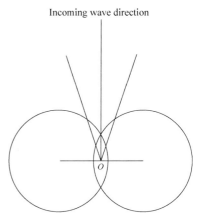

Angle Measurement Based on Angle Measuring Elements

The angle measuring elements include the induction synchronizer, rotary transformer, photoelectric coder, etc., which are mainly used in the stabilization platform,
stabilization gyrocompass, and gimballed inertial navigation systems.

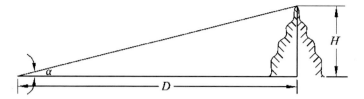

Fig. 4.37 Schematic diagram of the visual ranging

4.7.2 Range Measurement Method

Visual Range

This is the simplest ranging method. The distance from the carrier to the object is determined by the height of the known object (e.g., mountaintop or beacon) in the visible range (Fig. 4.37). The smaller Dis, the larger the angle α, and the higher the height H observed in the human eye. Commonly used measuring instruments are theodolite, telescopes with level, etc. This approach usually requires extensive experience, and the error is generally large. In practice, the distance of the optical measuring device and the ruler on the target point are determined, and the error is relatively small.

Homologous Bidirectional Ranging

This method adopts an active detection method. The signal source may emit electro-magnetic waves, acoustic waves, light waves or other media to reach the target and back to the signal source and calculates the distance of the target according to the signal propagation time.

According to the characteristics of waves propagating in a uniform medium, the distance r between two points is proportional to the wave propagation time t, then

$$r = \frac{t}{2}(v + u) \tag{4.37}$$

where u is the known wave velocity and v is the carrier velocity.

When laser, infrared and electromagnetic wave ranging is used, the carrier speed can be ignored. When acoustic wave ranging is used, the carrier speed cannot be ignored. Homologous bidirectional ranging can be divided into the following two categories according to the signal reversion mode.

(1) Target Reflection Mode

In this mode, the signal source produces a wave radiating to the target through direct passive reflection of the target, and the range is determined once the propagation time difference is obtained. Typical systems of this category include radar, sonar, photoelectric ranging systems, etc.

(2) Target Retransmission Mode

Different from the target passive reflection mode, the target reference point of this mode has the signal retransmission function, which can actively transmit the received signal back to the source. Actually, the ranging system consists of both the signal source and the target reference point. The retransmission mode can improve the action range of the system, improve the signal selection ability, and transmit other information through the channel. Typical systems of this category include Beidou I, the Loran system and the TACAN system.

Heterologous One-Way Range

This mode refers to the ranging method that directly transmits electromagnetic waves, acoustic waves and light waves from the signal source to the target. Without reflecting from the target, the relative target distance to the signal source can be measured by the one-way transmission time from the source to the target. To measure the propagation time of the wave, the target clock is required to maintain precise synchronization with the source clock. The ranging can be realized by measuring the propagation time difference or phase difference of the wave, etc.

1) Time Difference Ranging

This mode refers to the time interval between two RF signals with a pulse envelope, such as a pulse radio ranging navigation system, or by a pseudoranging method, such as a satellite positioning system:

$$R = c \cdot t \tag{4.38}$$

2) Phase Difference Ranging

The ranging is realized based on the phase of the received radio wave. The phase change in radio wave propagation is related to the propagation time t, i.e.,

$$\Phi = \omega \cdot t = 2\pi \frac{c}{\lambda} \cdot \frac{r}{c} = 2\pi \frac{r}{\lambda} \tag{4.39}$$

Then, we obtain

$$r = \frac{\lambda}{2\pi} \cdot \Phi \tag{4.40}$$

where Φ is the total phase delay caused by signal propagation; λ is the modulated signal wavelength; ω is the modulated signal angle frequency, and

$$\omega = 2\pi f = 2\pi \frac{c}{\lambda} \qquad (4.41)$$

When the wavelength (frequency) of the received signal is invariant, the phase difference is measured to obtain the distance. To measure the phase difference, the start of the phase reading also needs to be synchronized with the starting phase of the emitted wave; thus, a signal source with a very stable frequency is needed. Meanwhile, since the phase is a periodic function, there is a periodic multivalue problem in the phase measurement, which needs to be eliminated. Although these problems exist, the phase ranging measurement method is highly accurate and convenient, so it is widely used.

Multisource Distance Difference Ranges

This method adopts two transmitters of different base stations to transmit coordinated signals at the same time. The carrier receiver receives two signals at the same time and measures the transmission difference between the two signals through the equipment to obtain the distance difference.

Passive Range Measurement

This method measures the target distance by using the attenuation rule of the signal during the propagation of different channels, also known as the signal attenuation ranging method. If the original strength of the signal is known, the distance between the carrier and the signal transmitting source can be determined according to the strength of the received signal (RSS) received by the carrier, thus determining the position of the carrier itself. The general multiple of decay can be $1/r^2$ (r is the distance between the carrier and the signal source).

Appendix 4.1: Questions

1. Considering the general formula of direct positioning introduced in this chapter, please describe the characteristics and types of direct positioning.
2. Among the direct positioning methods such as two azimuth positioning, two horizontal angle positioning and azimuth-distance positioning, please choose one of them to derive an expression for its position function.
3. What is running fix? What are the conditions and key points when it is used? Please illustrate examples of its application in practice.

4. Please explain the basic principle of altitude difference celestial positioning, and point out the key to understanding this principle. Please briefly describe the general process, key points and application conditions of celestial positioning.
5. Please briefly described the positioning principle and characteristics of GPS, Glonass, Beidou and Galileo satellite navigation systems. Retrieve the information to illustrate the construction of these systems, Compare the positioning methods, characteristics and applications of different satellite navigation systems.
6. Please briefly describe the principle of radio navigation, and compares different positioning methods, characteristics and application fields of radio navigation system.
7. Please summarize the methods of indoor positioning commonly used. Illustrate the positioning principle of the mobile phone through data retrieval.
8. Please searching information about Malaysia Airlines MH370 passenger plane missing incident, as a world-class positioning problem, what is the main difficulty? Considering the maritime satellite "ping" signal and other information, please analyzes the various positioning technologies worldwide used for MH370 search.

Appendix 4.2: The Celestial Positioning Principle—"Knowing the Truth from the Untruth"

The basic principle of celestial positioning is based on the observation of the altitude angle of the stars. Since a position circle with an equal altitude angle can be obtained on the ground, multiple intersecting circles of equal altitude can be obtained through the observation of multiple stars, and the intersection point is the carrier position. However, the wide areas covered by the circles make it impossible to be drawn on a large-scale chart. In 1875, the French navigator Saint-Hillel proposed the altitude difference method, which solved this problem and made the celestial positioning method truly practical.

The idea is as follows: The bottleneck problem is how to draw the circle of equal altitude on the chart. Because the region displayed on the chart is much smaller compared to the Earth, the circular center, which is the substellar point of the star on the ground, usually falls outside the chart and thus cannot be drawn. Hillel suggested to use an imagined carrier position on the chart (e.g., the dead reckoning position) so that through geometric operation, the connection between the false position and the star substellar point can be calculated. Next, the circular radial line pointing to the star substellar point (the center of the circle) can be marked out. Then, the difference (altitude difference) between the observed altitude of the real position and the theoretic altitude of the false position can be calculated, where the altitude difference actually reflects the distance between the imagined point and the real point. Therefore, with the false point as a starting point, the vertical line can be drawn out along the altitude difference of the azimuth line. Since the region displayed on the chart is basically a point when compared to the substellar point, the altitude circle turns out to be a linear segment in the region of the chart. By so doing, the vertical line representing an arc of altitude circle, once difficult to draw, can be made by two altitude circles based on the observation of two stars. To put it simple, the intersection point of the two vertical lines on two different azimuth lines from the imaged point is the actual position of the celestial location of the carrier.

Chapter 5
Vector-Based Attitude Determination

Direction is the basic need of human beings. People need to identify the direction in their daily life. Direction is the basis of vehicle motion control, and identifying direction needs to establish the direction reference. How to establish the direction reference is a fundamental navigation problem.

The determination of attitude also has wide usage, such as artillery tube initial alignment, ship navigation attitude determination, directional antenna pitch control, and dam deformation monitoring. Attitude determination also needs to establish the level reference first.

There are many kinds of attitude determination methods. The simplest method is to find natural reference, such as the natural celestial bodies in the sky or obvious landmarks on the ground. Naturally observable celestial bodies in the sky can be used, such as the sun, moon, and stars. This method is called celestial direction-finding. Ground markers that use lighthouses, mountain tops or islands as references are called physiographic direction-finding. However, the two methods have certain limitations in use. The celestial direction-finding method is not usable under conditions of poor weather and low visibility due to the inability to see stars. The physiographic direction-finding method does not easily find obvious signs in some geographical environments, such as oceans or deserts. Therefore, the key is to establish or find the markers that can be used as a heading orientation reference. This idea continues to today. With the development of radio, optics, underwater acoustic and other measurement technologies, these reference objects can be radio stations, underwater acoustic beacons and other forms. Accordingly, a variety of different heading determination methods based on external references have been produced.

The simplest method of horizontal acquisition is obtained by observing the sea and sky boundary line or using a plane perpendicular to the plumb line based on gravity properties, and accordingly, attitude determination is achieved. Ship trim can be achieved by observing draught differences and other methods. With the development of technology, various kinds of attitude determination equipment and

© Science Press 2024
H. Bian et al., *Essentials of Navigation*, https://doi.org/10.1007/978-981-99-5636-4_5

instruments have emerged, such as level, theodolite for static measurement, longitudinal clinometer for dynamic inclination measurement, inertial attitude determination equipment, celestial (optical) attitude measuring equipment and satellite navigation attitude determination equipment for vehicle dynamic measurement.

Similar to the principle of positioning, attitude determination can also be divided into direct attitude determination and deduced attitude determination. Direct attitude determination can be divided into vector-based attitude determination and gimbal-based attitude determination. This chapter introduces the principle of direct attitude determination based on vector-observed, and the other two methods will be introduced in Chaps. 7 and 8.

5.1 Principle of Vector-Based Attitude Determination

The essence of acquiring the attitude information is to determine the angle relationship of the body coordinate system (b system) in the navigation coordinate system (n system). Vector-based attitude determination directly determines the heading and attitude of the vehicle by comparing the relationship between the vector angular measurements in the b system and the known angular values in the n system. Vectors are selected in various ways, including geophysical field vectors such as geomagnetic vectors, gravity vectors, earth rotation angle velocity vectors, or artificial construction vectors such as line-of-sight vectors and baseline vectors. According to the number of vectors adopted, it can be divided into the single-reference vector method, double-reference vector method and multiple-reference vector method.

5.1.1 Principle of Double Vector-Based Direct Attitude Determination

Solution of Attitude Matrix Based on Double Vectors

The heading and attitude parameters describe the angle relationship between the b system and the geographic coordinate system (g system). The double vector-based method determines the attitude of the vehicle by obtaining two noncollinear reference vectors simultaneously. This method can completely determine all attitude parameters of the vehicle in the navigation system.

Suppose $\boldsymbol{\Phi}_{\beta b} = \left(\varphi_{\beta\alpha}, \theta_{\beta\alpha}, \psi_{\beta\alpha}\right)$ is the attitude angle of coordinate system b relative to coordinate system β. If two noncollinear vectors $\boldsymbol{s}_1$ and $\boldsymbol{s}_2$ are known to have projections of $\boldsymbol{s}_1^{\beta}$ and $\boldsymbol{s}_2^{\beta}$ in the coordinate system β, and have projections of $\boldsymbol{s}_1^{b}$ and $\boldsymbol{s}_2^{b}$ in the coordinate system b, when both vectors are obtained by measurement or theoretical calculation, the third vector can be constructed. Note:

$$s_3^b = s_1^b \times s_2^b, s_3^\beta = s_1^\beta \times s_2^\beta \tag{5.1}$$

If the attitude matrix from the coordinate system b to the coordinate system β is C_α^β, then

$$s_i^\beta = C_b^\beta s_i^b \quad (i = 1, 2, 3) \tag{5.2}$$

The cosine matrix is solved by the equations:

$$C_b^\beta\left(s_1^b, s_2^b, s_3^b\right)\left(s_1^\beta, s_2^\beta, s_3^\beta\right)^{-1} \tag{5.3}$$

Since the reference vectors are noncollinear, the attitude matrix C_α^β can be uniquely determined.

Calculation of Attitude and Heading Angle Based on Attitude Matrix

According to Eqs. (2.25), (2.33), (2.35), (2.36), and (2.37), the attitude cosine matrix C_b^β and the attitude angle $\Phi_{\beta b}$ satisfy the following relationship:

$$C_\beta^b = \begin{bmatrix} \cos\theta_{\beta b}\sin\psi_{\beta b} & \cos\theta_{\beta b}\sin\psi_{\beta b} & -\sin\theta_{\beta b} \\ \begin{pmatrix} -\cos\phi_{\beta b}\sin\psi_{\beta b} \\ +\sin\phi_{\beta b}\sin\theta_{\beta b}\cos\psi_{\beta b} \end{pmatrix} & \begin{pmatrix} \cos\phi_{\beta b}\cos\psi_{\beta b} \\ +\sin\phi_{\beta b}\sin\theta_{\beta b}\sin\psi_{\beta b} \end{pmatrix} & \sin\phi_{\beta b}\cos\theta_{\beta b} \\ \begin{pmatrix} \sin\phi_{\beta b}\sin\psi_{\beta b} \\ +\cos\phi_{\beta b}\sin\theta_{\beta b}\cos\psi_{\beta b} \end{pmatrix} & \begin{pmatrix} -\sin\phi_{\beta b}\cos\psi_{\beta b} \\ +\cos\phi_{\beta b}\sin\theta_{\beta b}\sin\psi_{\beta b} \end{pmatrix} & \cos\phi_{\beta b}\cos\theta_{\beta b} \end{bmatrix} \tag{5.4}$$

Then, the attitude angle is calculated according to the following formula:

$$\begin{cases} \phi_{\beta b} = \arctan\dfrac{C_{\beta,2,3}^b}{C_{\beta,3,3}^b} \\[2ex] \theta_{\beta b} = \arctan C_{\beta,1,3}^b \\[2ex] \psi_{\beta b} = \arctan\dfrac{C_{\beta,1,2}^b}{C_{\beta,1,1}^b} \end{cases} \tag{5.5}$$

where arctan means the arctangent function and $C_{\beta,i,j}^\alpha$ denotes the element in the j-th column of the i-th row of the cosine matrix C_β^α.

This method which uses the transformation relation between matrices to obtain the relative attitude angle between coordinate systems b and β, is called the vector-based attitude determination principle.

5.1.2 Principle of Multiple Vector-Based Direct Attitude Determination

The multiple vector-based method uses more than two noncollinear vectors observed at each time to realize attitude determination. This method can not only determine the attitude of the carrier in the navigation coordinate system but also improve the calculation accuracy of the attitude matrix caused by suppressing measurement errors.

Assuming that the attitude sensor on the body obtains several noncollinear navigation vectors, the attitude matrix can be determined by the integration of these measurements. However, due to errors in vector measurements, different analyses of measurements will result in different attitude matrices. Therefore, when there is redundancy in measurement, there is a problem of how to obtain the optimal estimation of the attitude matrix. The least squares method is a common method to solve this problem.

Taking the navigation coordinate system n as the reference coordinate system β, where Formula (5.2):

$$\left(s_i^n\right)^T = \left(C_b^n s_i^b\right)^T = \left(s_i^b\right)^T\left(C_b^n\right)^T = \left(s_i^b\right)^T C_n^b \tag{5.6}$$

In the formula, T represents matrix transposition.

Let the measurement error be w_{si}, and the actual observation quantity $z_{s_i} = \left(s_i^n\right)^T$ satisfies:

$$z_{s_i} = \left(s_i^b\right)^T \cdot C_n^b + w_{s_i} \quad (i = 0, 1, 2 \cdots, m) \tag{5.7}$$

When multiple measurements are obtained, the following measurement equations can be obtained:

$$z_s = C_b^n \cdot s^b + C_b^n \cdot w_s \tag{5.8}$$

In the above formula:

$$z_s = \begin{bmatrix} z_{s_1} \\ z_{s_2} \\ \vdots \\ z_{s_m} \end{bmatrix}, \quad s^b = \begin{bmatrix} \left(s_1^b\right)^T \\ \left(s_2^b\right)^T \\ \vdots \\ \left(s_m^b\right)^T \end{bmatrix}, \quad w_s = \begin{bmatrix} w_{s_1} \\ w_{s_2} \\ \vdots \\ w_{s_m} \end{bmatrix} \tag{5.9}$$

where m is the number of observations.

The attitude matrix of the body can be obtained by solving the above equations by the least squares method.

$$\hat{C}_n^b = \left[(s^b)^T s^b\right]^{-1}(s^b)^T z_s \tag{5.10}$$

Then the attitude angle is calculated according to formula (5.5).

5.1.3 Characteristics of Direct Attitude Determination

Compared with the deduced attitude determination, navigation systems adopting direct attitude determination have the following characteristics:

(1) Relative attitude can be achieved. Because the vectors used can be measured directly in the navigation coordinate system and the body coordinate system, roll, pitch and heading can be calculated directly.

(2) Achieve fast attitude determination. The direct attitude determination system obtains real-time vector projection in the body coordinate system and navigation coordinate system by different measurement means. Compared with the deduced attitude determination, the direct attitude determination can achieve real-time and faster attitude determination.

(3) Accurate attitude determination can be achieved without accumulative error.

(4) Attitude determination error is related to vector selection and measurement error. Multiple sets of vectors are often selected for direct attitude determination, and their errors are mainly related to the selected vectors and their measurement errors.

(5) Availability is greatly affected by motion and the environment. Direct attitude determination generally requires good stability and horizontal reference of the vehicle because good meteorological and observation conditions need external reference.

5.2 Principle of Geomagnetic Heading and Attitude Determination

The geomagnetic field is a vector field (northward component, eastward component, vertical component and its combined total field vector) that is continuously distributed in the global near-Earth space. Geomagnetic field intensity is a function of geographic position and time and is mainly composed of magnetic field components with different variations. It provides a natural coordinate reference system for spaceflight, aviation and navigation and is widely used in positioning, orientation and attitude control of spacecraft and naval vessels.

Geomagnetic attitude determination is an accurate, reliable and efficient orientation technology that is valuable in applications. A magnetic heading measuring instrument is also called a magnetic compass, which uses the earth magnetic field to measure the direction of the device. An electronic magnetic compass that can output electrical signals is composed of magnetic sensors, electronic circuits and microprocessors. It is widely used in the automation system of aircraft, tanks, ships and other vehicles.

It should be noted in particular that there is a geomagnetic declination between geomagnetic north and geographic north because the Earth's rotation axis does not coincide with the geomagnetic axis. Geomagnetic declination is an important parameter of the geomagnetic field and a traditional spatial geographic element loaded by military maps and charts. The magnetic heading or magnetic azimuth measured by the geomagnetic sensor must be corrected by the magnetic declination angle to obtain the "real" vehicle geographic heading or azimuth. Geomagnetic navigation technology has many advantages, such as simplicity, reliability, anti-interference, and low cost. Accurate calculation of geomagnetic declination must be based on high-precision geomagnetic field observation data. Marine geomagnetic declination maps are an important basis for submarine navigation underwater [1, 5].

5.2.1 Basic Principles of Geomagnetic Heading and Attitude Determination

Flux Density of the Local Geomagnetic Field

The magnetic flux density of the earth's magnetic field is represented by subscript E, which can be expressed in the local navigation coordinate system.

$$m_E^n(p_b, t) = \begin{pmatrix} \cos\alpha_{nE}(p_b, t)\cos\gamma_{nE}(p_b, t) \\ \sin\alpha_{nE}(p_b, t)\cos\gamma_{nE}(p_b, t) \\ \sin\gamma_{nE}(p_b, t) \end{pmatrix} B_E(p_b, t) \qquad (5.11)$$

In the formula, B_E is the amplitude of the magnetic flux density; α_{nE} is the magnetic deflection angle or magnetic change; γ_{nE} is the magnetic inclination angle of the geomagnetic field; and all three parameters are functions of position and time (Fig. 5.1).

The magnetic flux inclination angle is essentially the geomagnetic latitude, so it is within $10°$ of the Earth's latitude L_b. The magnetic deflection angle gives the direction of the geomagnetic field relative to true north, which is the only one of the three parameters for the geomagnetic field orientation. The magnetic deflection angle can be regarded as a function of position and time, which can be calculated using global models such as the 275 coefficient international geomagnetic reference field (IGRF) or 336 coefficient US/UK world magnetic model (WMM).

Measurement of Total Flux Density by Magnetometer

The magnetometer measures the total flux density, expressed by subscript m. If the magnetometer sensitivity axis is aligned with the inertial sensor sensitivity axis, and the measured value of the magnetometer is decomposed into the body coordinate

Fig. 5.1 Schematic diagram of the magnetic flux density vector

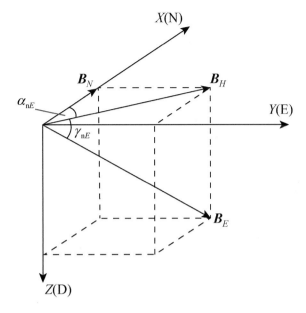

system b and can be expressed as

$$m_m^b = C_n^b \begin{pmatrix} \cos\alpha_{nm}\cos\gamma_{nm} \\ \sin\alpha_{nm}\cos\gamma_{nm} \\ \sin\gamma_{nm} \end{pmatrix} B_m \tag{5.12}$$

In the formula, B_m, α_{nm} and γ_{nm} are the amplitude, deflection and inclination of the total flux density, respectively.

Principle of Double Vector-Based Geomagnetic Attitude Determination

From the principle formula of vector-based attitude determination (5.2)

$$m_m^b = \begin{pmatrix} \cos\theta_{nb} & 0 & -\sin\theta_{nb} \\ \sin\phi_{nb}\sin\theta_{nb} & -\cos\phi_{nb} & \sin\phi_{nb}\cos\theta_{nb} \\ \cos\phi_{nb}\sin\theta_{nb} & \sin\phi_{nb} & \cos\phi_{nb}\cos\theta_{nb} \end{pmatrix} \begin{pmatrix} \cos\psi_{mb}\cos\gamma_{nm} \\ \sin\psi_{mb}\cos\gamma_{nm} \\ \sin\gamma_{nm} \end{pmatrix} B_m \tag{5.13}$$

In the formula, ϕ_{nb} is the rolling angle; θ_{nb} is the pitch angle; and Ψ_{mb} is the magnetic heading angle, given by the following formula:

$$\psi_{mb} = \psi_{nb} - \alpha_{nm} \tag{5.14}$$

Types of Geomagnetic Heading and Attitude Determination Systems

According to whether a stabilizer is adopted, the geomagnetic heading and attitude determination system can be divided into two categories: stabilizer type and non stabilizer type.

The rolling angle and pitch angle can be approximately zero with a stabilizer. The magnetic heading measurement can be obtained by the magnetometer through the following formula:

$$\tilde{\psi}_{mb} = \arctan \frac{-\tilde{m}^b_{m,y}}{\tilde{m}^b_{m,x}} \tag{5.15}$$

If there is no stabilizer, the rolling angle and pitch angle are nonzero and need to be measured by MEMS sensors or other inertial measurement methods. Magnetic heading and attitude determinations are:

$$\tilde{\psi}_{mb} = \arctan \frac{-\tilde{m}^b_{m,y}\cos\hat{\phi}_{nb} + \tilde{m}^b_{m,z}\sin\hat{\phi}_{nb}}{\tilde{m}^b_{m,x}\cos\hat{\theta}_{nb} + \tilde{m}^b_{m,y}\sin\hat{\phi}_{nb}\sin\hat{\theta}_{nb} + \tilde{m}^b_{m,z}\cos\hat{\phi}_{nb}\sin\hat{\theta}_{nb}} \tag{5.16}$$

5.2.2 Environmental Magnetic Field Effect on Geomagnetic Heading Finding

Unlike other heading and attitude determination techniques, the difficulty of the geomagnetic method is that the magnetic force not only measures the geomagnetic field and local anomalies but also the magnetic field of the navigation system itself, the vehicle and the environment. The magnetic properties of equipment can be divided into hard ferromagnetism and soft ferromagnetism. Hard ferromagnetism is a magnetic field produced by permanent magnet iron and electronic equipment, while soft ferromagnetism is generated by the distortion of the underlying magnetic field caused by the material and only under the effect of an external magnetic field. Soft ferromagnetism is larger on ships than on aircraft and land vehicles.

The total flux density measured by the magnetometer assembly is

$$m^b_m = b_m + (I_3 + M_m)C^b_n(m^n_E + m^n_A) + w_m \tag{5.17}$$

where m^n_E is the flux density of the geomagnetic field; m^n_A is the abnormal flux density of the local magnetic field; b_m is the component of the hard iron flux density in the body coordinate system; M_m is the scale factor and cross coupling matrix of soft iron; and w_m is the random magnetic measurement noise of the magnetometer.

The magnetometer has zero bias, scale and cross-coupling errors. The magnetism of the equipment is in the relative body coordinate system, while the environmental magnetism is relative to the earth. Thus, the magnetic flux density of the equipment and the environment can be distinguished. Considering that the azimuth, rolling and pitch of the magnetic compass will change during swinging, $\boldsymbol{b}_m$ and $\boldsymbol{M}_m$ can be calibrated by a series of measurements in the swing process. Because the position is fixed, the ambient flux density is assumed to be constant. Therefore, the calibration parameters and ambient magnetic flux density can be estimated using the nonlinear estimation algorithm built in the magnetic compass. The magnetometer measurements are then compensated with the following formula:

$$\hat{\boldsymbol{m}}_m^b = \left(\boldsymbol{I}_3 + \hat{\boldsymbol{M}}_m\right)^{-1}\tilde{\boldsymbol{m}}_m^b - \hat{\boldsymbol{b}}_m \tag{5.18}$$

In the formula, $\hat{\boldsymbol{b}}_m$ and $\hat{\boldsymbol{M}}_m$ are hard ferromagnetism and soft ferromagnetism estimated.

5.2.3 Magnetic Compass

The magnetic compass is a basic navigational instrument. The sensitive element of a magnetic compass is a magnetic needle. The geomagnetic horizontal force H is in the north direction of the horizontal magnetic field, which makes the magnetic needle point to the north horizontally. It is called the magnetic north force. The magnetic north force is strongest at the magnetic equator and gradually weakens with increasing latitude, reaching zero at the magnetic north and south poles. Therefore, when ships sail to high-latitude areas, the magnetic compass pointing ability becomes weaker, its accuracy is poorer, and it cannot work normally near the magnetic pole.

A magnetic compass is a geomagnetic heading-finding device with a stabilizer. To improve the sensitivity of the magnetic compass, it is necessary to reduce the friction force of magnetic needle rotation. As shown in Figs. 5.2 and 5.3, by injecting liquid into the compass bowl, the compass disc is suspended on the axis needle, thus counteracting the influence of gravity and minimizing the friction force.

Principle of Self-Deviation

Due to the use of iron and steel materials in ships, an inductive magnetic field will be produced under the influence of the earth's large magnetic field. Obviously, the magnetic needle of the magnetic compass is affected not only by the earth magnetic field but also by the influence of the ship's own magnetic field, resulting in the deviation of the direction of the magnetic compass. As shown in Fig. 5.4, the magnetic north (N_m) of the earth is not in the same direction as the true north (N_t). The angle between the magnetic north and the true north is called the magnetic difference (Var). The angle between compass north (N_c) and magnetic north (N_m) is called the compass deviation, expressed as δ or Dev. The self-deviation of the compass is positive to the east and negative to the west.

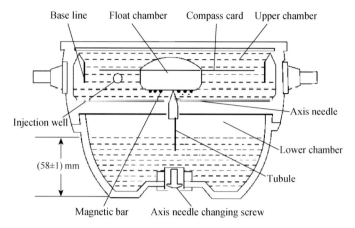

Fig. 5.2 Structure of the compass bowl, with permission from reference [China Science Publishing & Media Ltd. (CSPM)]

Fig. 5.3 Compass bowl, photograph by Bian H.W.

Fig. 5.4 Magnetic difference and deviation

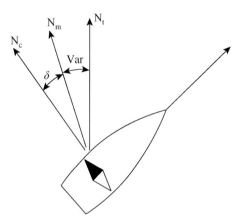

The ship's hard iron magnetic field originated in the period of ship construction. Because the ship direction is fixed for a long time during builting, the ship hard iron is magnetized under the influence of the earth magnetic field. The magnitude and polarity of the magnetism are fixed relative to the ship. A soft iron magnetic field is only produced under the influence of an external magnetic field, and its magnitude and direction are changed as the external magnetic field changes. Therefore, the magnetic needle in the compass bowl is actually affected by the geomagnetic, hard iron magnetic, and soft iron magnetic forces.

For magnetic compasses with an approximately constant level controller, a simple four coefficient calibration can be achieved by simply swinging the compass around the heading axis. Heading correction can be achieved by using the following formula:

$$\hat{\psi}_{mb} = \tilde{\psi}_{mb} + \delta\hat{\psi}_{mb} \tag{5.19}$$

Note that this calibration is effective only when the magnetic compass is horizontal.

The deviation formula is:

$$\delta\hat{\psi}_{mb} = \hat{C}_0 + \delta\hat{C}_{b1} + \delta\hat{C}_{b2} + \delta\hat{C}_{s1} + \delta\hat{C}_{s2} \tag{5.20}$$

where $\delta\hat{C}_{b1} = \hat{C}_{b1}\sin\tilde{\psi}_{mb}$, $\delta\hat{C}_{b2} = \hat{C}_{b2}\cos\tilde{\psi}_{mb}$, $\delta\hat{C}_{s1} = \hat{C}_{s1}\sin2\tilde{\psi}_{mb}$, and $\delta\hat{C}_{s2} = \hat{C}_{s2}\cos2\tilde{\psi}_{mb}$.

In the formula, $\hat{C}_{b1}$ and $\hat{C}_{b2}$ are hard ferromagnetic correction coefficients, $\hat{C}_{s1}$ and $\hat{C}_{s2}$ are soft iron correction coefficients, $\hat{C}_0$ does not change with the course M_C, called constant self-deviation; $\delta\hat{C}_{b1}$ and $\delta\hat{C}_{b2}$ change twice in one circle, called semicircular self-deviation; and $\delta\hat{C}_{s1}$ and $\delta\hat{C}_{s2}$ change 4 times in one circle, called quadrant self-deviation.

Elimination of Self-Deviation

The principle of eliminating self-deviation is to use the same magnitude, opposite direction and the same nature of magnetic force to offset the ship magnetic force. Hard ferromagnetism is eliminated with a magnetic rod. Soft iron magnetism is eliminated by soft iron bars.

(1) semicircular self-deviation

The Airy method is usually used; that is, the self-deviation is observed directly on the 4 basis points heading 0°, 90°, 180° and 270°, and the semicircular self-deviation is offset by a horizontal longitudinal and horizontal correction magnet.

Calibration step:

- Sailing along the magnetic north (N); measure out the error δ_N; adjust the transverse magnetic rod to eliminate δ_N.
- Sailing along the magnetic east (E) or weat (W); measure out the error δ_E (or δ_W); adjust longitudinal magnetic rods to eliminate δ_E (or δ_W).
- Sailing along the magnetic south (S); measure out the error δ_S; adjust the transverse magnetic rod to eliminate half of δ_S; $\delta\hat{C}_{b2}$ is cancelled.
- Sailing along the magnetic west (W) or east (E), measure the error δ_W (or δ_E), and adjust the longitudinal magnetic rod to eliminate half of δ_W (or δ_E); $\delta\hat{C}_{b1}$ is cancelled.
- Quadrant self-deviation

Normally, δ_E is small enough to be negligible. For $\delta\hat{C}_{s1}$:

- Sailing along the heading northwest (NW), measure out the error δ_{NW}, move soft iron bars to eliminate δ_{NW};
- Sailing along the heading southwest (SW), measure out the error δ_{SW}, move soft iron bars to eliminate half of δ_{NW}; cancel $\delta\hat{C}_{s1}$
- Calculating the self-deviation coefficient

When compensating for the self-deviation, it is impossible to eliminate all self-deviation entirely, and there is always residual self-deviation left. It is necessary to determine the coefficients of the self-deviation formula of the magnetic compass and then make the self-deviation table or draw the curve of the self-deviation for ship navigation.

- Measure the residual self-deviations δ of magnetic compasses in eight different headings, such as N, NE, E, SE, S, SW, W and NW. The residual self-deviations are substituted into the self-deviation formula (5.19).
- Eight formulas containing approximate self-deviations are obtained, and the approximate self-deviation coefficients $\hat{C}_0$, $\delta\hat{C}_{b1}$, $\delta\hat{C}_{b2}$, $\delta\hat{C}_{s1}$, $\delta\hat{C}_{s2}$ are calculated and substituted. The self-deviation formula change with the heading value $\hat{\psi}_{mb}$ is obtained.
- According to the self-deviation formula, calculate the self-deviations every $10°$ or $15°$, and fill in the results in the self-difference form. Plot the relationship curve between the calculated self-deviation and the magnetic compass heading. The ordinate is the self-deviation, and the abscissa is the magnetic compass heading. In normal operation, the self-deviation value corresponding to the current magnetic compass heading is found directly from the curve, and the self-deviations are eliminated by compensating for the modified magnetic compass indication (Figs. 5.5 and 5.6).

Fig. 5.5 Self-deviation table

0°	2.0	180°	2.0
10°	1.9	190°	1.9
20°	1.5	200°	1.5
30°	1.0	210°	1.0
40°	0.3	220°	0.3
45°	0.0	225°	0.0
50°	−0.3	230°	−0.3
60°	−1.0	240°	−1.0
70°	−1.5	250°	−1.5
80°	−1.9	260°	−1.9
90°	−2.0	270°	−2.0
100°	−1.9	280°	−1.9
110°	−1.5	290°	−1.5
120°	−1.0	300°	−1.0
130°	−0.3	310°	−0.3
135°	0.0	315°	0.0
140°	0.3	320°	0.3
150°	1.0	330°	1.0
160°	1.5	340°	1.5
170°	1.9	350°	1.9

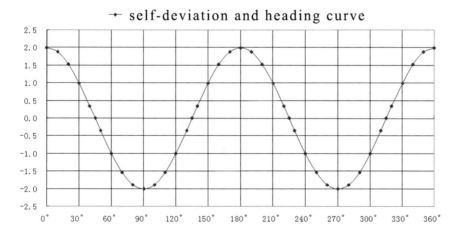

Fig. 5.6 Self-deviation and heading curve

5.3   Principle of Celestial Attitude Determination

The principle of celestial positioning is introduced in Chapter IV. In fact, there is a long history of identifying azimuth information through celestial objects. It is the most traditional, direct azimuth-finding method. Today, celestial attitude determination can directly obtain high-precision carrier attitude information, which is an important means of long-term navigation for aircraft, ships and spacecraft. The high-precision attitude information provided by the celestial navigation attitude system can be used to estimate and correct the attitude error of the carrier, which is also the basis of celestial navigation and positioning.

5.3.1   Measurement Principle of Star Sensor

The basic principle of celestial attitude determination is to measure the attitude of the vehicle relative to the geocentric inertial coordinate system (ECI) by the star sensor [18]. To obtain the attitude information of the vehicle, the key is to determine the attitude of the body coordinate system in the navigation coordinate system. The star sensor is the most accurate attitude sensor on spacecraft thus far. It plays an irreplaceable role in high-precision mapping, remote sensing and formation flying. Whether in earth orbit satellites or deep space detectors, large space structures or small satellites, star sensors are almost always used to determine high-precision attitude. Compared with other attitude sensors, star sensors can provide angular measurement accuracy of arc-second or even sub-arc second and can realize three-axis absolute attitude determination of spacecraft [17]. The navigation stars referenced by the star sensor are relatively uniform throughout the celestial sphere, so a star sensor with appropriate sensitivity and field of view can detect stars in almost any direction, which is incomparable with other attitude sensors [1, 5].

Star sensors mainly include two parts: a sensitive system and a data processing system. The sensitive system consists of shade, an optical lens and a sensitive surface array, which mainly achieves the acquisition of sky navigation star map data. The data processing system achieves the processing of acquired navigation star map data and the determination of attitude. The composition of the star sensor is shown in Fig. 5.7.

As shown in Fig. 5.8, the star sensor uses the principle of aperture imaging to measure the star characteristics of a navigation star f and the line of sight vector relative to the body coordinate system b:

$$\boldsymbol{u}_{bf}^b = \frac{1}{\sqrt{x_{cf}^{c2} + y_{cf}^{c2} + F^2}} \boldsymbol{C}_c^b \begin{pmatrix} -x_{cf}^c \\ -y_{cf}^c \\ F \end{pmatrix} \tag{5.21}$$

Fig. 5.7 Composition of
star sensor

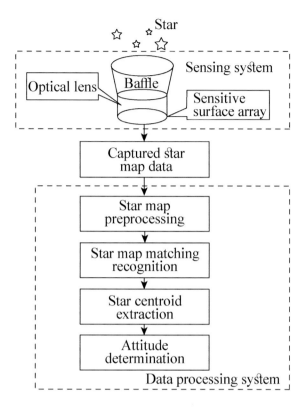

Fig. 5.8 Imaging principle
of star sensor

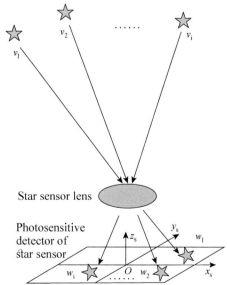

where $\left(x_{cf}^c, y_{cf}^c \right)$ is the position coordinate of the star sensor spindle center on the detector; F is the focal length of the star sensor; and C_c^b is the known star sensor to the body coordinate system.

In fact, a more accurate model is needed to consider the optical deformation and other systematic errors of the star sensor lens. The line of sight vector is related to azimuth ψ_{bu}^{bf} and elevation θ_{bu}^{bf} by

$$
u_{bf}^b = \begin{bmatrix} \cos \theta_{bu}^{bf} \cos \psi_{bu}^{bf} \\ \cos \theta_{bu}^{bf} \sin \psi_{bu}^{bf} \\ -\sin \theta_{bu}^{bf} \end{bmatrix}
$$

$$
\theta_{bu}^{bf} = -\arcsin\left(u_{bf,z}^b\right)
$$

$$
\psi_{bu}^{bf} = \arctan_2\left(u_{bf,y}^b, u_{bf,x}^b\right)
$$
(5.22)

5.3.2 Principle of Multivector Celestial Attitude Determination

At present, the angular position accuracy of navigation stars in the star list is generally on the order of 20 ms. Compared with other errors of star sensors, its accuracy can be considered sufficiently high. Stars that satisfy the imaging conditions of star sensors are selected from the star database to form a navigation star list. The navigation list is recorded in the memory of the star sensor when on the ground [19].

After many years of celestial observations, the position of the astral features of each star f is relatively fixed in the celestial sphere (Fig. 5.9), generally expressed in the right ascension a_{if} and declination δ_{if} in the sphere coordinates of the celestial sphere. Then, the projection of the line-of-sight vector in the ECI inertial coordinate system is expressed as

$$
u_{if}^i = \begin{pmatrix} \cos \delta_{if} \cos \alpha_{if} \\ \cos \delta_{if} \sin \alpha_{if} \\ \sin \delta_{if} \end{pmatrix}
$$
(5.23)

where u_{if}^i has the following relation with the line-of-sight vector u_{bf}^b observed by the star sensor in the center of the star image:

$$
u_{if}^i = C_b^i u_{bf}^b, \quad f \in 1, 2 \cdots
$$
(5.24)

where C_b^i is the transformation matrix from the body coordinate system to the inertial coordinate system.

Fig. 5.9 The star in celestial
coordinate system and
rectangular coordinate
system

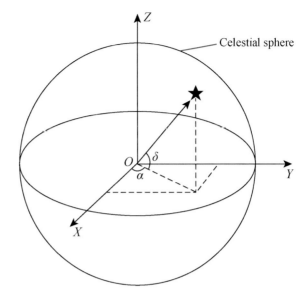

According to the principle of double-vector attitude determination, more than
two star line of sight vectors are required to solve $\boldsymbol{C}_b^i$. When more than 2 stars are
observed, the least squares estimation can be used to determine the optimal attitude
solution. Nearly 10 stars are commonly used in practical applications.

The attitude matrix of the vehicle relative to the local navigation coordinate system
can be expressed as:

$$\boldsymbol{C}_b^n = \boldsymbol{C}_e^n \boldsymbol{C}_i^e \boldsymbol{C}_b^i \tag{5.25}$$

If the time, latitude and longitude of the vehicle are known, then:

$$\boldsymbol{C}_i^e = \begin{bmatrix} \cos(\omega_{ie}t) & \sin(\omega_{ie}t) & 0 \\ -\sin(\omega_{ie}t) & \cos(\omega_{ie}t) & 0 \\ 0 & 0 & 1 \end{bmatrix} \tag{5.26}$$

$$\boldsymbol{C}_e^n = \begin{bmatrix} -\sin\lambda_b & \cos\lambda_b & 0 \\ -\sin\lambda_b\cos\lambda_b & -\sin L_b\sin\lambda_b & \cos L_b \\ \cos L_b\cos\lambda_b & \cos L_b\sin\lambda_b & \sin\lambda_b \end{bmatrix} \tag{5.27}$$

5.3.3 Celestial Positioning Principle Based on Known Attitude

The position of the vehicle can also be calculated when the roll and pitch are provided to the celestial navigation system by inertial navigation or tilt sensors in other navigation systems. From the ECEF coordinate system to the navigation coordinate system, the transition matrix is given by the following formula:

$$C_e^n = C_b^n C_i^b C_e^i \tag{5.28}$$

where

$$C_b^n = \begin{bmatrix} \cos\theta_{nb}\cos\psi_{nb} & \begin{pmatrix} -\cos\phi_{nb}\sin\psi_{nb} \\ +\sin\phi_{nb}\sin\theta_{nb}\cos\psi_{nb} \end{pmatrix} & \begin{pmatrix} \sin\phi_{nb}\sin\psi_{nb} \\ +\cos\phi_{nb}\sin\theta_{nb}\cos\psi_{nb} \end{pmatrix} \\ \cos\theta_{nb}\sin\psi_{nb} & \begin{pmatrix} \cos\phi_{nb}\cos\psi_{nb} \\ +\sin\phi_{nb}\sin\theta_{nb}\sin\psi_{nb} \end{pmatrix} & \begin{pmatrix} -\sin\phi_{nb}\cos\psi_{nb} \\ +\cos\phi_{nb}\sin\theta_{nb}\sin\psi_{nb} \end{pmatrix} \\ -\sin\theta_{nb} & \sin\phi_{nb}\cos\theta_{nb} & \cos\phi_{nb}\cos\theta_{nb} \end{bmatrix} \tag{5.29}$$

The latitude and longitude expressions are as follows:

$$L_b = -\arcsin\left(C_{e,3,3}^n\right)$$
$$\lambda_b = -arctam\left(-C_{e,3,2}^n, -C_{e,3,1}^n\right) \tag{5.30}$$

5.3.4 Marine Celestial Navigation Equipment

Traditional celestial navigation should be based on an accurate nautical almanac, accurate measurement time, altitude of the star, horizontal reference and probable position of the observer. At present, both the nautical almanac and the time have been well obtained. The key problem is the measurement of the star altitude and the level reference establishment of the observer.

Before photoelectric and inertial stabilization techniques are applied to celestial positioning, it mainly uses the sea and sky boundary line as the horizontal reference, so only in good weather conditions can the observer measure the star altitude. Due to the influence of weather and observation methods, traditional methods cannot achieve real-time and accurate positioning.

Today's marine celestial navigation systems are generally used in combination with inertial navigation systems, attaching the star tracker to an inertial platform and forming a celestial/inertial integrated navigation system. Through the optimal estimation and compensation for the errors of the inertial navigation system, the integrated system can work independently without relying on other navigation means and output the position, heading, attitude, speed and other navigation information.

The marine celestial navigation system is mainly composed of a star tracker, display console, power controller and other components. Among them, the star tracker is the main piece of equipment. The international typical celestial navigation equipment is the daystar celestial navigation equipment (American Microcosm Inc.), with a star measurement capacity of magnitude +7 in the daytime, can be used for ships and aircraft. Daytime Stellar Imager celestial navigation equipment (American Trex Enterprises Corporation), also known as the optical GPS system, is capable of observing magnitude +6 stars during the daytime. In November 2006, the Northrop Grumman company completed the development and test work of the LN120G stellar-inertial navigation system, which can track stars day and night with 20″ orientation precision and optimize the location and direction information of the inertial navigation system.

Celestial navigation is widely used in the fields of marine, aviation and aerospace. As an effective navigation method, the main features of celestial navigation are as follows.

(1) Passive measurement and autonomous navigation. Celestial navigation takes the celestial body as the navigation beacon, does not rely on other external information, does not radiate energy to the outside, passively receives the celestial body radiation radio wave or light and electrical signals, and then obtains navigation information.

(2) High navigation accuracy without error accumulation. The accuracy of celestial navigation mainly depends on the accuracy of the celestial sensitivity and reaches the order of angular seconds. It provides the most accurate direction information that can be used as the calibration reference of inertial navigation equipment. The error does not accumulate with time, which is more important for the carrier of long-time conceal maneuvers.

(3) Strong anti-interference ability and high reliability. The celestial body radiation covers the entire electromagnetic wave segment of X-ray, ultraviolet, visible light and infrared light and has a very strong anti-interference ability. The spatial motion of celestial bodies cannot be disturbed by humans.

(4) Provide various navigation information. Celestial navigation can provide the position and attitude information of the carrier, as well as the speed and other information.

(5) Affected by the external environment. In the field of aviation and maritime, celestial navigation is susceptible to meteorological conditions, and the system will not be used when the star cannot be observed normally.

5.4   Principles of GNSS Attitude Determination

Attitude determination technology based on GNSS has high precision and low cost and can provide accurate attitude information without divergence over time. GNSS carrier phase measurement technology is mainly achieved by measuring the change in the carrier phase, whose accuracy is much higher than that of code-based pseudorange measurement, and it is not limited by the secrecy of the precision code (p code or y code). As a result, carrier phase measurement technology has become the main method for GNSS high-precision positioning and attitude determination.

5.4.1   Principle of Carrier Phase Measurement

The distance (ρ) between the GNSS receiver and the satellite can be expressed either by the transmission time τ of the signal multiplied by the speed of light c, that is, $\rho = c \cdot \tau$. It can also be represented by the phase difference of the signal carrier:

$$\rho = \lambda(\varphi_S - \varphi_R) + N\lambda \qquad (5.31)$$

In the formula, φ_S represents the phase of the carrier signal at satellite S; φ_R is the phase of the carrier signal at receiver R; λ is the wavelength of the carrier signal; and N is the integer number of the period of the carrier signal propagation.

However, because the receiver operates in a passive state, the carrier phase at satellite φ_S cannot be measured. The problem can be solved by simulating the carrier signal in practice. When the carrier loop of the user receiver locks the satellite carrier signal, the oscillator in the receiver produces a signal with the same frequency and initial phase as the satellite carrier signal, which is called the reference signal. Therefore, ($\varphi_S - \varphi_R$) in formula (5.31) is equal to the phase difference between the reference signal phase generated by the receiver and the received satellite carrier signal phase, that is,

$$\varphi_s - \varphi_R = \varphi(\tau_b) - \varphi(\tau_a) \qquad (5.32)$$

where $\varphi(\tau_b)$ is the reference signal phase when the receiver receives the signal at time τ_R and $\varphi(\tau_a)$ is the phase of the satellite carrier signal measured by the receiver at transmission time τ_S.

This is the general principle of the carrier phase measurement. The actual carrier phase observations (expressed in $\tilde{\varphi}$) consist of the integer part of the period number $Int(\phi)$ and the decimal part $F(\phi)$ less than one period, as shown in Fig. 5.10. The $Int(\phi)$ of the first observation is zero, and the $Int(\phi)$ of the subsequent continuous observation at t_i is obtained by Doppler continuous counting, which can be a positive integer or a negative integer. Since the carrier is simply a cosine wave without any identification marks, there is an entire period unknown number N_0 between $\tilde{\varphi}$ and

Fig. 5.10 Carrier phase observation

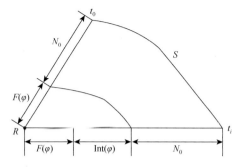

Fig. 5.11 GNSS compass, photograph by Bian H.W

$\varphi(\tau_b) - \varphi(\tau_a)$ of each observation, which is called the initial integral period ambiguity N_0.

5.4.2 Basic Principles of Satellite Navigation Attitude Determination

Satellite navigation position measurement adopts a multi-vector attitude determination mode to realize the carrier attitude determination. A pair of antennas (Fig. 5.11) simultaneously fixed to the carrier measures the distance between the satellite and the two antennas simultaneously using the GNSS carrier phase, and the relative positioning results can be used to obtain the attitude information of the carrier. Since the baseline distance between the two antennas is much less than the distance from the antenna to the satellite, the vector of sight between the pair of antennas to a given satellite can be regarded as parallel [1, 5]. Therefore, the angle θ between the baseline and the line of sight vector can be given by the following equation:

Fig. 5.12 GNSS attitude
determination

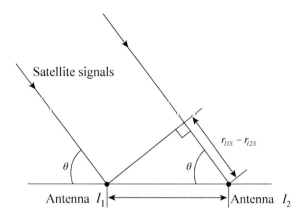

$$\cos\theta = \frac{r_{l_1 S} - r_{l_2 S}}{r_{l_1 l_2}} \tag{5.33}$$

where $r_{l_1 S}$ and $r_{l_2 S}$ are the distances between satellite S and antennas l_1 and l_2, respectively. $r_{l_1 l_2}$ is the known distance between the two antennas, as shown in Fig. 5.12. When the receiver position is known, the line-of-sight vector from the satellite to the antenna is known, so the attitude information of the vehicle relative to the earth can be obtained:

$$\tilde{r}^n_{l_1 l_2} = \tilde{C}^n_e r^e_{l_1 l_2} \tag{5.34}$$

After obtaining the measurement information of the baseline $\tilde{r}^n_{l_1 l_2}$ in the local navigation coordinate system, the following equation is associated with the known carrier coordinate system baseline vector $\tilde{r}^b_{l_1 l_2}$:

$$r^n_{l_1 l_2} = \tilde{C}^n_b r^b_{l_1 l_2} \tag{5.35}$$

A unique solution of attitude C^n_b cannot be obtained based on a single vector. Only two elements can be obtained from a single baseline measurement. To solve this problem, a third antenna (expressed in l_3) that is noncollinear with two existing antennas must be introduced to establish a second baseline, having

$$\tilde{r}^n_{l_3 l_2} = \tilde{C}^n_b r^b_{l_3 l_2} \tag{5.36}$$

Construct the third vector according to (5.1), which is given:

$$\tilde{r}^n_{l_1 l_2} \times \tilde{r}^n_{l_3 l_2} = \tilde{C}^n_b (r^b_{l_1 l_2} \times r^b_{l_3 l_2}) \tag{5.37}$$

According to (5.3), the attitude result is obtained:

$$\tilde{C}_b^n = (\tilde{r}_{l1l2}^n, \tilde{r}_{l3l2}^n, \tilde{r}_{l1l2}^n \times \tilde{r}_{l3l2}^n) \cdot (r_{l1l2}^b, r_{l3l2}^b, r_{l1l2}^b \times r_{l3l2}^n)^{-1} \qquad (5.38)$$

More antennas can be used to obtain more accurate carrier attitude according to the multi-vector attitude determination principle. This technique is also known as interferometric attitude determination or GNSS compass.

5.4.3 Analysis of Satellite Attitude Determination Accuracy

The attitude determination accuracy can be given by the ratio between the carrier phase measurement accuracy of the baseline and the baseline length. Therefore, for a 1 m rigid baseline, if the measurement accuracy is 1 cm, then the standard deviation of the attitude determination is 10 mrad (approximately 0.6°). The longer the baseline is, the higher the attitude determination accuracy. However, the longer the baseline is, the more difficult it is to meet the rigidity condition of the baseline in the dynamic environment. The elastic deformation of the baseline will reduce the accuracy.

Satellite navigation attitude determination has many advantages: ① high accuracy, better than the traditional magnetic compass and gyrocompass; ② low cost, significantly lower than the above attitude determination equipment; ③ simple structure, simple maintenance. Therefore, it is widely used in both civilian maritime and aviation fields.

In addition, because the heading accuracy of the satellite attitude determination system does not have the latitude-related gyrocompass effect of inertial attitude equipment, it can still work normally in high latitude areas and indicate the carrier heading accurately, so it is designated by the International Maritime Organization (IMO) as the necessary attitude determination equipment for ships in the polar area. The disadvantage of satellite navigation attitude determination is that it depends on the normal reception of satellite signals. When the satellite signals are blocked or disturbed, the system will not work normally, so it needs to be combined with other technical means.

Appendix 5.1: Questions

1. Write down the double-vector direct positioning formula, explain its principle and process.
2. Considering the formula and principle of multi-vector direct positioning, please explain the main features and application of vector observation.
3. From the perspective of the principle of the vector-based attitude determination, please illustrate the basic principle of the geomagnetic attitude determination, and key points of the geomagnetic compass difference calibration.
4. Please briefly describe the basic principle of celestial attitude determination and the relation with celestial positioning.

5. Briefly describe the basic principle of multi-antenna satellite attitude difference and collect information about relevant products and applications.

6. Comparing the direct positioning general formula and the vector-based attitude determination formula, please analyze the connection and difference between them.

7. Retrieve and analyze various new attitude determination methods are analyze the relation and differences between them and the vector-based attitude determination.

Appendix 5.2: The simple theory that is closer to the essence of the problem

"Induction, comparison, and connection establishment" are important scientific ways of thinking. Take Maxwell's equations for example. Through the thoughtful comparison and analysis of different theorems (i.e., Gauss' theorem, Biosaval's law, and Faraday's law), Maxwell established the connections among them and eventually unified them into Maxwell's equations, which not only laid the foundation of Electromagnetism but also successfully predicted the existence of electromagnetic waves. "Induction, comparison, and connection establishment" can help simplify the questions and the differences among them, and have the power of innovative thinking to guide basic or application innovation.

In the same way, we could embrace these scientific ways of thinking when learning different positioning modes (e.g., celestial positioning, satellite positioning, and radio positioning) and attitude-determining modes (e.g., celestial attitude determination, satellite attitude determination, and geomagnectic attitude determination). Despite the differences among the different modes, induction and comparison help us identify their commonalities, and then better understand the differences.

In summary, the minimalist way of thinking is efficient in helping people quickly grasp the essence of complex problems. In practice, people may not be able to get the point at first, but the minimalist mindset will help them to think and explore more actively. As an old Chinese saying goes, "the greatest truths are the simplest (大道至简)", which also reflects an attitude to pursuing the law of things. People should be encouraged to reinforce this efficient way of thinking.

Chapter 6
Velocity Measurement

Velocity is a fundamental parameter of carrier motion. Velocity measurement has a very important role in the navigation control, position calculation and route planning of the carrier. Velocity is a common basic physical quantity that is closely related to many other physical quantities. Its measurement devices are various, such as the log on the ship, the pitot tubes on the aircraft, and the odometer on the car. Velocity and distance are inseparable. In the process of dealing with practical problems, if the distance of the object displacement is known, then its speed is easily obtained by measuring the moving time. If the object velocity is known, then the position of a certain period of time can also be determined.

6.1 Main Speed Measurement Principle

The old-fashioned sector log lowers the knotted rope from the stern of the ship and then determines the ship's speed by counting the knots of the rope, which is why the knot is used as a common speed unit in navigation. Currently, many different kinds of logs have been developed, such as electromagnetic logs, Doppler logs and acoustic correlation logs. Among them, sonar is an important navigation instrument that is used to measure speed. An odometer is usually used to measure the speed in land vehicles by measuring the rotation of wheels. This technology can be traced back to the ancient Han Dynasty (202 BC–220 AD) and is widely used for all land vehicles at present. Currently, pedestrian walking distance can be counted automatically by a pedometer or by a more accurate pedestrian dead reckoning (PDR) method based on an accelerometer that can calculate the number and size of steps. The flight speed of aircraft can be determined by the Doppler frequency shift of the ground echo wave of radar or by comparing continuous terrain images tracked by camera, radar or laser scanner.

© Science Press 2024

H. Bian et al., *Essentials of Navigation*, https://doi.org/10.1007/978-981-99-5636-4_6

There are many methods of velocity measurement that can be roughly divided into three groups: the kinematics principle, dynamic control principle and velocity-related physical principle.

6.1.1   Velocity Solution based on Kinematic Principles

The velocity solution based on the principle of kinematics is a mechanical method that studies the object position change rule over time from the geometric point of view (does not involve the physical stress of the object). Kinematic studies examine motion features such as the equation of motion, displacement, velocity and acceleration. Therefore, velocity measurements can be obtained based on the principle of kinematics.

Calculate the Speed based on the Displacement

The relationship between displacement ΔS, time Δt and velocity of the vehicle is as follows:

$$V = \frac{\Delta S}{\Delta t} \tag{6.1}$$

For the known and fixed displacement, the passing time needs to be measured. For the known and fixed time periods, the position of observation points at the beginning and end of time needs to be measured. The displacement and time should be measured simultaneously. Here are some simple examples of these applications:

Example. 1 Transit beacon method

The so-called transit beacon in navigation refers to 2 to 3 sets of transit beacons with a special top mark established on the shore. The transit beacon lines are parallel to each other, and the vertical distance between the transit beacon lines is generally 1, 2 or 3 nautical miles (Fig. 6.1). The starting and ending points of the route are observed by transit beacons, and the starting time t_1 and the ending time t_2 are recorded by a stopwatch. Then, the real speed is obtained by the known vertical distance divided by the time interval recorded.

The main method of using transit beacons to measure ship speed is to keep the ship's speed unchanged during speed measurement, and the ship sails in a course perpendicular to the transit beacon. The ship speed can be obtained by the formula (unit of speed V: kn; unit of time t: s).

$$V = \frac{S}{t_2 - t_1} \times 3600 \tag{6.2}$$

Fig. 6.1 ship velocity measurement by transit beacons

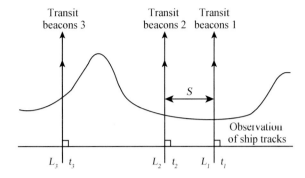

Example. 2 Vehicle Speed Measurement

To test the safety performance of automobiles, automobile manufacturers have to carry out impact tests in which the speed of automobiles needs to be measured outside. A common method is to mark the surface of the vehicle. The distance between these marks is known. The speed of the vehicle can be calculated by measuring the time difference between different marks using optical sensors to record the time of vehicle marks passing through.

Example. 3 GNSS Velocity Measurement

Since the GNSS can provide precise position and time, the ship speed can also be measured if this information is used correctly. First, choose the appropriate sea area and time for testing, keep the ship running at a constant speed, and use the GNSS receiver to record the position and time of the test starting point and ending point. Sample multiple sets of point position and time recorded and calculate ship speed according to formulas (6.3) and (6.4) (unit of speed V: kn, unit of distance ΔS: nautical mile).

When $\varphi_S \neq \varphi_E$,

$$\Delta S = \left| \frac{\varphi_S - \varphi_E}{\cos\left[\tan^{-1}\left(\frac{\lambda_S - \lambda_E}{\varphi_S - \varphi_E} \cos\frac{\varphi_S - \varphi_E}{2}\right)\right]} \right| \text{(unit:nautical mile)} \tag{6.3}$$

When $\varphi_S \approx \varphi_E$,

$$\Delta S = \left| (\lambda_S - \lambda_E) \cos\frac{\varphi_S + \varphi_E}{2} \right| \text{(unit:nautical mile)} \tag{6.4}$$

Example. 4 Odometer Speed Measurement

An odometer is a device used to record the distance of a car, which is based on the output data of the transmission system, such as the output axle of the gearbox or the wheels. Based on the speed of the wheel (e.g., the number of rotations per minute), the known circumference of the wheel (i.e., the distance the wheel travels in one circle) and the fixed time Δt, the vehicle mileage Δs is calculated. By dividing the mileage of the car by the fixed time, the speed of the car during this period can be obtained.

Calculating Speed based on the Integration of Acceleration

Acceleration is the ratio of the change of velocity with respect to time $\Delta v / \Delta t$, using symbol a to express and the unit is m/s^2. Acceleration is a vector quantity (it has magnitude and direction). Velocity can be obtained by integrating acceleration based on self-measured acceleration. When acceleration over a period of time $a(t)$ is known, the expression of average velocity in a certain period of time is as follows:

$$V(t) = \int_{t_0}^{t} a(t)dt + V(t_0) \tag{6.5}$$

where t_0 is the initial moment of motion, $V(t_0)$ is the initial velocity of the object, and $V(t)$ is the average velocity of the object during this period of time.

Measuring the velocity of an object by its acceleration is a common method in inertial navigation. The inertial navigation system uses an accelerometer to directly obtain the acceleration of the object in a certain direction to calculate the velocity of the vehicle. The inertial navigation velocity measurement will be further introduced in Chap. 8.

6.1.2   *Velocity Measurement Based on Propulsion System*

The power system is simply a machine that provides power, such as the engine of an automobile or a ship. The car engine provides power to turn the wheels, which drives the car to move. We know that the speed of a car can be roughly calculated by the rotation speed of the wheel multiplied by its circumference. Set the wheel rotation speed (numbers of turns per time unit) to be N, the diameter of the wheel is L, and then the vehicle speed V is as follows:

$$V = N\pi L \tag{6.6}$$

In the situation of ships, because the propeller is driven by the transmitter and the main engine, the main engine speed cannot indicate the speed accurately but only reflects the speed of the ship. The measurement of linear motion is transformed into the measurement of rotational change by measuring the velocity of the power system, which is an approximate method of velocity measurement.

6.1.3 Velocity Measurement Based on the Physical Effect

There are many physical effects related to velocity. For example, the Bernoulli effect, Faraday effect, Doppler effect and so on. According to these physical effects, various kinds of speed measuring devices can be designed. These include the pit log (Bernoulli effect), electromagnetic log (Faraday effect), and Doppler velocity log (Doppler effect). On the ship, the accumulated mileage of the ship is calculated by time integration based on various kinds of logs. When further combined with heading measurement equipment (e.g., gyrocompass), it can also estimate the real-time position of the vehicle when the starting point is known. The principle of velocity measurement based on physical effects is the main method of carrier velocity measurement and will be introduced in the following parts.

6.2 Bernoulli Effect-Based Velocity Measurement

There are many kinds of equipment to measure velocity by the Bernoulli principle, and the most common application is the pit log of ships or the pitot tubes of aircraft. The Bernoulli principle was proposed in 1726 by Daniel Bernoulli, a Dutch mathematician and physicist. The essence of the Bernoulli principle is the conservation of mechanical energy of fluid, namely:

Kinetic energy + gravitational potential energy + pressure potential energy = constant.

According to Bernoulli's principle, the sum of kinetic energy, potential energy and pressure at each section of the same streamline remains unchanged when the ideal fluid flows steadily. The equation is expressed as follows:

$$\frac{1}{2}mv^2 + mgh + pV = C \tag{6.7}$$

where:

m—Quality of liquid flowing through the cross section
v—Velocity of liquid flow
g—Gravity acceleration
h—Height of liquid cross section relative to datum level
p—Pressure on the cross section of liquid
V—Volume of liquid flowing through the cross section
Let $\gamma = \frac{mg}{V}$, γ be the specific gravity of liquids, and the formula is changed to:

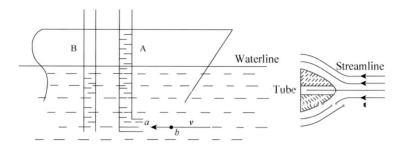

Fig. 6.2   The schematic diagram of the principle of speed measurement of pit log

$$\frac{v^2}{2g} + h + \frac{1}{\gamma}p = C \tag{6.8}$$

Generally, assuming sea level to be horizontal, the flow of sea water relative to the ships can be regarded as horizontal flow. To measure the ship's speed relative to sea water, two tubes A and B are projected below its water line (Fig. 6.2). The mouth of tube A is toward the bow, facing the flow direction, and its section is vertical to the direction of ship motion; the mouth of tube B is downward, and its section is parallel to the direction of ship motion. When a ship is in the stationary state relative to sea water, that is, the relative velocity of ship motion $V = 0$, the pressures of A and B are equal, and they are both under hydrostatic pressure. The hydrostatic pressure is proportional to the draft depth of the ship. When the ship moves at a speed relative to the sea water of v, the mouth of tube B is not affected by water flow because it is parallel to the direction of water flow, and it is still under hydrostatic pressure, so tube B is called the hydrostatic pressure tube. Because tube A is facing the water flow, the pressure of tube A is not only static pressure but also dynamic pressure produced by sea water, so tube A is called a full pressure tube.

Choose a point a at the mouth center of the full pressure tube and take a very tiny stream parallel to the ship's draft line; the velocity of the tiny stream is equal to that of the velocity of the sea water relative to the ship's motion v. Choose a point b at any place on this streamline, using P_a, P_b, v_a, v_b to express the pressure and velocity of a and b, respectively. The Bernoulli equation at two points a and b is written as below.

$$\frac{v_a^2}{2g} + h + \frac{p_a}{\gamma} = \frac{v_a^2}{2g} + h + \frac{p_b}{\gamma} \tag{6.9}$$

At point a, because the water flow is obstructed by the water inside tube A, it cannot keep flowing forward, and $v_a = 0$. At point b, because the water flow is unimpeded, its velocity is still equal to the velocity of the ship relative to the sea water, namely, $v_b = v$. Substitute v_a and v_b in the formula above and simplify:

$$P_a = \frac{\gamma}{2g}v^2 + P_b \tag{6.10}$$

In the above formula, P_b is the static pressure at water depth h, P_a is the resulting pressure of the static pressure P_b and the dynamic pressure P_d:

$$P_d = P_a - P_b = \frac{\gamma}{2g} v^2 \tag{6.11}$$

Formula (6.10) can be abbreviated as (6.11)

$$v = \sqrt{\frac{2 P_d}{\rho}} \tag{6.12}$$

where $\rho = \gamma/g$.

When the density of water ρ is constant, the dynamic pressure P_d is only proportional to the square of the speed. Therefore, as long as the difference between dynamic and static pressure is measured, the ship's speed can be calculated.

6.2.1 Pit Log

Pit log is based on Bernoulli's principle. It can indicate the speed and distance of the ship by using the principle that the dynamic pressure is proportional to the square of the ship's speed relative to the current. The difference between dynamic and static pressure is measured by a differential pressure sensor, and the velocity of the ship relative to the sea water is calculated.

The most typical device for measuring dynamic pressure is a hydraulic system consisting of a pitot tube, full pressure tube and hydraulic box (Fig. 6.3). At the bottom of the ship, a hydrostatic pressure tube and full pressure tube were installed, through which sea water flowed into the hydraulic box. The hydraulic box is installed in the cabin below the water line, and its internal diaphragm is divided into two chambers: the upper chamber and the lower chamber. The upper chamber is connected to the hydrostatic tube, and the static pressure from the orifice of the hydrostatic tube is transmitted through the hydrostatic pipeline and then applied to the diaphragm from above. The lower chamber is connected to the full pressure tube, and the static and dynamic pressure from the orifice of the full pressure pipeline are applied to the diaphragm from below. When the distance of the hydrostatic tube orifice and the full pressure orifice are equal to the sea surface, the hydrostatic pressure of sea water to both orifices are almost equal. Therefore, the hydrostatic pressures on the upper and lower diaphragms are equal. After cancelling the hydrostatic pressure, only the dynamic pressure transmitted by the full pressure pipeline is left, which makes the diaphragm deform upward, produces output, and drives the pointer to indicate the speed.

The characteristics of the pit log are listed as follows:

(1) Reliable performance;
(2) Good accuracy at medium and high speeds;
(3) Poor accuracy and sensitivity at low speed;

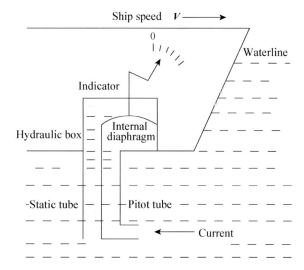

Fig. 6.3 Schematic diagram of pit log

(4) Tubes of the hydraulic system are extended out of the bottom of the ship at a length of 300–900 mm, which makes the ships pick up them when sailing through the fishing areas or entering and leaving ports.

With the development of sensing technology, a new generation of pit logs adopts intelligent differential pressure sensors to measure dynamic pressure, which greatly improves the accuracy and reliability.

6.2.2   Aircraft Speedometer

Airspeed is the forward component of the aircraft velocity relative to air. The airspeed can be obtained by measuring the pressure difference between the forward-mounted pitot tube (measuring the dynamic pressure) and the side-mounted pitot tube (measuring the static pressure).

It should be noted that both the pit log and the **aircraft speedometer** measure relative velocities. The speed measured by the pit log is not the speed of the ship relative to the seabed but the speed relative to the sea water, which is classified as the relative log. When there is a current, the real speed of the ship relative to the seabed needs to be added to the current (sailing along with the current) or subtracted from the current (sailing against the current). In addition, the measured velocity error is also related to the density of water. The airspeed has a similar feature. Attention should be given to speed calculation and processing of speed errors.

Fig. 6.4 Faraday's law of
electromagnetic induction

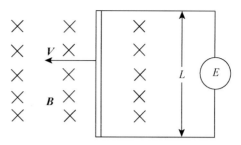

6.3 Faraday Effect-Based Velocity Measurement

6.3.1 Faraday Effect

According to Faraday's law of electromagnetic induction, when there is a relative velocity V between the conductor and a uniform magnetic field with a magnetic induction intensity of B, the conductor cuts the magnetic curve and then produces an inductive potential E at both ends (Fig. 6.4). If the effective length of the conductor is L and three kinds of vectors (B, L, V) are perpendicular to each other, then:

$$E = BLV \tag{6.13}$$

When the magnetic field is an alternating field $B = B_m \sin \omega t$, the resulting inductive potential e will change in the same rule as B:

$$e = B_m LV \sin \omega t = E_m \sin \omega t \tag{6.14}$$

According to (6.14), we have $V = \frac{E_m}{B_m L}$.

The amplitude E_m of the induced electromotive force (EMF) is proportional to the velocity V. As long as the amplitude of the induced electromotive force is measured, the velocity of the conductor can be obtained.

6.3.2 Electromagnetic Log

Sea water is a kind of continuous conductive medium. When an alternating magnetic field is established in such a medium, if there is relative motion between the magnetic field and the medium, an induced potential will be generated in any closed loop through the medium.

The electromagnetic log measures the ship's speed according to the relationship between the induced potential generated by the moving conductor itself and the moving velocity in the magnetic field. Electromagnetic logs usually use electromagnetic sensors to measure the induced EMF. It consists of two parts: excitation part and induction electrode part. The excitation part produces an alternating magnetic

Fig. 6.5 Planar sensor

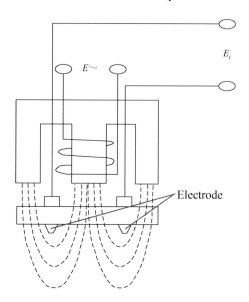

field, while the induction electrode part detects an inductive potential proportional to the velocity. The commonly used sensors are plane type and rod type.

Figure 6.5 shows the planar-type sensor. The excitation coil is wound around the mountain-shaped core, the induction electrode at the bottom is immersed in seawater, and the induction signal is drawn from the wire. The whole sensor is installed on the bottom of the ship, which is level with the bottom of the ship, and the line between the two electrodes is perpendicular to the keel line of the ship.

Figure 6.6 shows the rod-type sensor, which consists of a measuring rod and the measuring head. The excitation coil is wound around the strip core inside the measuring head and covered with a glass fiber layer. There is an induction electrode on the left and right sides, and the centerline between the two electrodes is perpendicular to the keel line of the ship. When working, the measuring head is extended to a certain distance from the bottom of the ship by a lifting mechanism, usually approximately 350–500 mm.

When the instrument works, the sensor excitation coil generates an alternating magnetic field along the bottom of the ship, and the sea water between the two electrodes acts as a conductor. When the ship moves, the magnetic curve at the bottom of the ship is cut by the sea water between the two electrodes, and the induced potential between the two electrodes is proportional to the speed.

The electromagnetic log uses the electromagnetic induction principle to measure ship speed and calculates the accumulative ship range. Similar to the pit log, the velocity measured is not the velocity of the ship relative to the seabed but relative to the water.

Its advantages are as follows:

(1) The sensitive inductive potential of the electromagnetic sensor is linearly related to the speed. It has not only high-speed measurement sensitivity but also a wide range of speed measurements.

Fig. 6.6 Measuring rod
sensor

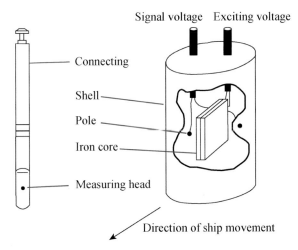

(2) It can also measure the ship's backward speed.
(3) The equipment is not affected by hydrological conditions such as density, temperature, salinity, pressure and conductivity in the water area.
(4) The inductive potential is generated instantaneously, which can reflect the change in the ship's instantaneous speed, and the speed measurement accuracy is high.

   The disadvantages are as follows:

(1) The sensor's long-term immersion in seawater will lead to a decline in performance due to the growth of sea creatures.
(2) The rod-type sensor will also need to be picked up in shallow waters and fishing areas to prevent damage.

   In other words, electromagnetic logs are widely used instruments.

6.4   Doppler Effect-Based Velocity Measurement

6.4.1   Doppler Effect

In 1842, Christian Doppler, an Austrian physicist, discovered that if there is relative motion between the source transmitting a fixed frequency signal and the receiver, the frequency of the signal received by the receiver will be different from the frequency emitted by the source. When the source and receiver are close to each other, the received frequency increases, and vice versa.

   We often find these phenomena in our daily life. For example, as a train whistles and rushes across the station, the sound of the whistle heard by people at the station

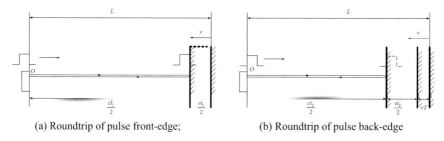

(a) Roundtrip of pulse front-edge;   (b) Roundtrip of pulse back-edge

Fig. 6.7 Doppler effect analysis in the time domain. **a** Roundtrip of pulse front-edge; **b** Roundtrip of pulse back-edge

varies with the distance between him and the train. As the train approached, the observer heard the whistle becoming sharper, and its frequency increased; as the train passed by and went away, the whistle sounded gradually changed from high to low, and its frequency decreased. This is the Doppler effect.

Now, the Doppler effect is analyzed in the time domain by taking the pulse signal emitted by a sound source as an example to obtain a general expression suitable for arbitrary pulse waveforms.

Consider the case in which the receiver and transmitter are at the same position. The target moves toward the transducer at speed v. The distance between the transducer and the target is L at the initial time $t = 0$, and the sound speed is c. The transmitter emits a signal with a pulse width T and frequency f to the target (Fig. 6.7). The time when the pulse front reaches the target and is reflected back to the receiving point is t_1, and the distance that the target approaches the sound source is vt_1 within t_1, see Fig. 6.7a; therefore,

$$L = \frac{vt_1}{2} + \frac{ct_1}{2} \tag{6.15}$$

Therefore, the round-trip time of the pulse front is

$$t_1 = \frac{2L}{c + v} \tag{6.16}$$

When the back edge of the pulse leaves the transducer surface, the target approaches the transducer for a distance vT (see Fig. 6.7b). If its round trip time is t_2, when the back-edge of the pulse reaches the target, the distance that the target approaches the transducer is vt_2. Thus:

$$L = vT + \frac{vt_2}{2} + \frac{ct_2}{2} \tag{6.17}$$

From this, the following can be deduced:

$$t_2 = \frac{2(L - vT)}{c + v} \qquad (6.18)$$

From formulas (6.16) and (6.18), we can see that the round trip times of the pulse front edge and back edge are different:

$$\Delta t = t_1 - t_2 = \frac{2vT}{c + v} \qquad (6.19)$$

Therefore, the pulse width of the received signal T_r is:

$$T_r = T - \Delta t = \frac{c - v}{c + v} T \qquad (6.20)$$

This indicates that when there is a radial relative motion between the target and the transducer, the original pulse signal will be compressed or widened. If the period of the original pulse signal is T, the period of the receiving pulse signal T_r is changed. Thus, the frequency of the received signal f_r is:

$$f_r = \frac{1}{T_r} = \frac{c + v}{c - v} \frac{1}{T} = \frac{c + v}{c - v} f = \frac{1 + x}{1 - x} f \qquad (6.21)$$

where f is the frequency of the transmitted signal and v is the sum of the radial velocity of the source and the target when both of them are moving.

6.4.2 Principle of the Doppler Velocity Log

Doppler velocity log (DVL) measures ship speed based on the Doppler effect of ultrasound waves propagating in water [20]. A transceiver is installed at the bottom of the ship to transmit an ultrasonic pulse signal with a frequency f_0 at the pitch angle θ to the bottom of the ship and to receive an echo signal reflected back through the seabed (Fig. 6.8).

If the ship moves at velocity V, it can be decomposed into component $V \cos \theta$ along the emission direction and component $V \sin \theta$ perpendicular to the emission direction. The $V \sin \theta$ component does not produce a Doppler effect, while the $V \cos \theta$ component moves the sound source (transducer) relative to the receiving point (at the seabed reflection). With the ship as the center, the relative motion speed of the seabed against the ship is $V \cos \theta$. According to Eq. (6.21), the frequency of the echo signal received at the receiving transducer is:

$$f_r = f_0 \frac{c + V \cos \theta}{c - V \cos \theta} \qquad (6.22)$$

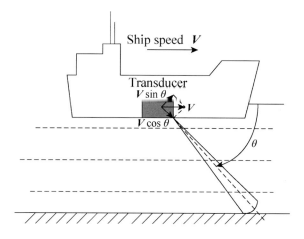

Fig. 6.8 Basic principle of doppler log

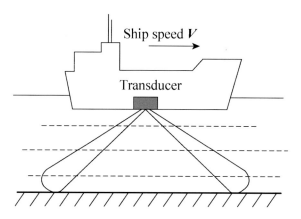

Fig. 6.9 Velocity measurement by dual wave

In the formula, c is the speed of ultrasound propagation in seawater; θ is the angle between the midline of the beam and the sea level called the beam dip angle; V is the ship velocity; and f_0 is the ultrasonic emission frequency.

Obviously, there is a Doppler shift between the transmitted signal and the received echo signal:

$$\Delta f = f_r - f_0 = 2f_0 \frac{V \cos\theta}{c - V \cos\theta} \tag{6.23}$$

Velocity component of sound source along beam propagation direction caused by ship motion $V \cos\theta \ll C$, so:

$$\Delta f = 2f_0 \frac{V \cos\theta}{c} \tag{6.24}$$

Or:

$$V = \frac{c}{2f_0 \cos \theta} \Delta f \tag{6.25}$$

As long as the Doppler shift Δf is measured, the ship's speed V can be obtained. This is the basis for measuring ship-to-ground velocity by Doppler log. However, when the wave around the ship is strong enough to make the vertical velocity of the ship large enough, the actual propagation velocity of the sound source in the ultrasonic emission direction is $V \cos \theta - U \sin \theta$ due to the influence of the vertical velocity U. The resulting Doppler shift Δf_F is therefore:

$$\Delta f_F = 2f_0 \frac{V \cos \theta - U \sin \theta}{c} \tag{6.26}$$

At this time, it would be inaccurate to calculate the speed according to formula (6.25). To eliminate the influence of the ship's vertical velocity on velocity measurement, the dual beam method is usually used. That is, a transducer is added to transmit ultrasonic waves to the bottom of the stern at the same frequency and pitch angle, as shown in Fig. 6.9. The Doppler shift obtained from it is as follows:

$$\Delta f_A = 2f_0 \frac{-V \cos \theta - U \sin \theta}{c} \tag{6.27}$$

The difference between the two transducers is calculated by using the Doppler frequency shifts obtained from the forward and backward transducers.

$$\Delta f' = \Delta f_F - \Delta f_A = 4f_0 \frac{V \cos \theta}{c} \tag{6.28}$$

We can see from the above formula that after adopting the dual beam method, has a linear relationship with the speed V and has nothing to do with the vertical speed U. At the same time, the frequency shift $\Delta f'$ is twice as large as that of a single beam, which is helpful to improve the sensitivity of the instrument. When the ship

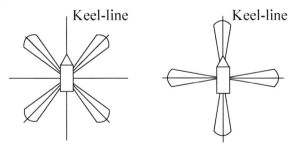

(a)X shape configuration   (b)Cross shape configuration

Fig. 6.10 Four-beam configuration

sways and tilts with two beams, the Doppler frequency shift sensed by the front and rear transducers is not exactly the same, but most of them can be cancelled. Thus, the influence of the vertical velocity on the velocity measurement can also be decreased. Therefore, in practice, the Doppler log adopts the dual-beam method to measure the ship's speed.

The features of the Doppler log are as follows:

(1) Because the Doppler log measures the velocity of a ship relative to the seabed based on the Doppler effect of sound waves, it will be affected by the hydrological conditions.

(2) When the depth of sea water exceeds hundreds of meters, sound waves are mainly reflected by mass particles in the ocean water layer. At this time, it measures the ship-to-water velocity. Therefore, the Doppler log can work in two ways: water layer tracking and seabed tracking.

(3) The Doppler log has high accuracy and a small measuring threshold, usually 0.01 kn.

(4) When two more beams are added to the dual beam type, i.e., the four-beam Janus type configuration (Fig. 6.11). The system can not only measure the longitudinal forward and backward velocities of ships but also measure the transverse velocities of ships. It is especially suitable for large ships to enter and leave narrow channels, dock or anchor with extremely low speed to the ground to ensure the safety of ship maneuvering.

(5) On the other hand, the Doppler log needs to radiate energy outward, which has a negative impact on the concealment of ships.

Fig. 6.11 Doppler log, with permission from reference [CSSC Marine Technology]

6.5 Acoustic Correlation-Based Velocity Measurement

An acoustic correlation log is used to measure the ship speed by the relationship between the correlation delay time and the speed of the underwater acoustic signal. It has the same accuracy as Doppler logs, but it is not affected by the speed of ultrasonic wave propagation in water; that is, it is not affected by changes in sea water temperature, salinity and depth [21].

Three transducers are installed at the bottom of the ship along the keel lines (Fig. 6.12), and from the bow to the stern in turn are receiving transducer 1, the transmitting transducer and receiving transducer 2. The transmitting transducer continuously emits ultrasonic signals to the seabed along the vertical direction. Receiver transducers 1 and 2 receive echo signals returned from the seabed (as shown in Fig. 6.13). The envelope amplitude of the echo signal is related to physical conditions such as sea water depth, seabed sediment, scattering ability of various scatterers in ultrasonic propagation, and absorption ability of seawater media to ultrasonic waves. Because of the different propagation paths, the envelope amplitudes of the echo signals obtained by receiving transducers 1 and 2 are not necessarily the same at a certain time, but there is a temporal correlation between them.

Fig. 6.12 Working principle of Acoustic Correlation-Based Log

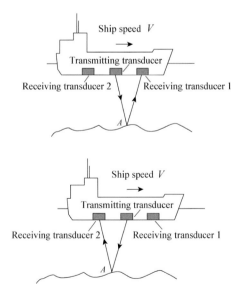

Fig. 6.13 Related delay
time of signals

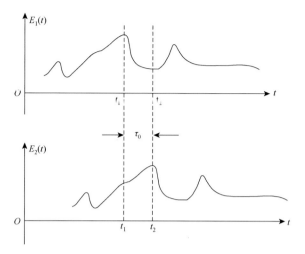

Set the ship to sail at speed V. At a certain moment (time t_1), the receiving transducer 1 receives the echo signal reflected by point A on the subsea. After time τ_0 (time t_2), the transmitting transducer moves to the position where the receiving transducer 1 is at time t_1, and the receiving transducer 2 moves to the position where the transmitting transducer is at time t_1. At this time, receiving transducer 2 will also receive the echo signal reflected by point A on the subsea. Since the two echo signals pass through the same path, but only in the opposite direction, the resulting envelope amplitude should be equal. The output of the receiving transducers 1 and 2 will produce a time difference of τ_0. The function curves of the signal amplitude envelopes $E_1(t)$ and $E_2(t)$ varying with time are almost identical (as shown in Fig. 6.11), so the two signals are interrelated.

If the installation distance between receiving transducers 1 and 2 is L, the relationship is as follows:

$$\tau_0 = \frac{L}{2V} \ or \ V = \frac{L}{2\tau_0} \tag{6.29}$$

L in the formula is a fixed value. If the delay time τ_0 can be measured, the speed can be obtained. The delay time τ_0 can be measured by correlation techniques.

The features of the acoustic correlation log are as follows:

(1) The accuracy of the acoustic correlation log is not affected by the velocity of acoustic wave propagation in seawater, that is, not affected by the temperature and salinity of seawater, so it is used to process the information of water depth to measure ship speed and accumulative range.
(2) It can either work in the "water layer tracking" mode as a relative log or work in the "seabed tracking" mode as an absolute log.
(3) The log has good linearity and high accuracy, the minimum speed measuring threshold is approximately 0.01 kn, and the range accuracy is nearly 0.2%.

(4) It is characterized by the wide diffusion angle of the transducer, which can
     reduce the leakage of the echo signal when the ship is sailing at high speed.
(5) The acoustic wave of the acoustic correlation log is transmitted vertically, so
     the depth of the transducer to the seabed can also be calculated by the time
     difference between transmitting and receiving, that is, it has the function of a
     deep sounder.

Appendix 6.1: Questions

1. Please retrieve and illustrate the ancient and modern speed measurement tech-
    nology to deepen the understanding of the historical development of speed
    measurement.
2. What are the practical application requirements of speed information? Please
    briefly describe the important role of speed information in navigation?
3. What are the differences in the speed parameters required for the route planning
    and transfer alignment processes? Analyze the causes.
4. Considering the content of this chapter, we briefly describe the classification and
    related application characteristics of the speed measurement method, such as
    speed measurement by automobile speed meters. Think about other speed-related
    principles and formulas in physics, which can they be applied to navigation?
5. Please retrieve and illustrate the principle and application of Doppler radar, and
    compare the similarities and differences with Doppler velocity log.
6. Briefly describe the velocity measurement formula of the pit log, and the types
    and characteristics of the speed measurement device based on the Bernoulli
    principle.
7. Briefly describe the speed measurement formulas of the EM log, DVL and
    acoustic correlation log, and their working principle and characteristics.
8. How can we understand the navigation culture of various countries from the
    perspective of navigation?

Appendix 6.2: Time thinking in ancient China

China has been attaching great importance to time since ancient times. For example,
the term "Sheng Chen Ba Zi" (birth date and the eight characters of a horoscope)
indicates the date and time of one's birthday by sexagenary cycle method. A total of
eight characters using celestial stems ("天干 Tian Gan") and terrestrial branches ("
地支Di Zhi") to represent the year, month, day and time of one's birth respectively.
It used to hold an important position as an old folk belief in China, and now it is often
regarded as oriental mysticism or some old superstition. Today with the navigation

knowledge we could understand the concept of time in ancient China more fairly and comprehensively.

The year, month, day, and time represented by eight characters indicate the position of the earth in the solar system (year), the position of the earth in the revolution orbit relative to the sun (month), the position of the earth relative to the moon (day), and the angular position of the Earth's rotation relative to the sun (time). Therefore, the term denotes a concept in the domain of modern astronomical time.

It will not be difficult for readers to understand the corresponding position relationship between the time concept ("month, day, time") and the celestial objects ("sun, earth, month") in space, but the concept of "year" turns out to be more complicated. Ancient Chinese called Jupiter the year-indicating star, and its revolution period is approximately 12 years (or 11.8618 years). The public regression period of the neighbor planets of the Earth, such as Jupiter and Saturn, is approximately 60 years (or 59.555 years). The movements of Venus, Jupiter, Mercury, Mars, and Saturn relative to the Earth also have cycles, although more complex. The celestial stems and terrestrial branches, or Tian Gan Di Zhi, are associated with periodic changes of celestial bodies in the position relative to the Earth.

As early as the Xia and the Shang dynasties, the ancient Chinese had the sense that "time" and "position" had important influences on the development of things. From the perspective of navigation, the eight characters of birth date reflect the position of the observer in the universe at the birth moment. Behind the position is the combination of position-related factors, such as gravity, magnetic field, light, and temperature, and the like, which in turn affect many extended influences of nature (such as seasons, tides, geological movements, etc.), biology (growth, maturity, aging, etc.), society, and psychology, like the ripples in the lake, layer by layer, stir each other, and pass on indefinitely. Therefore, the interaction of these effects is not merely multidimensional or nonlinear, and the reasons behind are more often than not difficult to clarify and understand. Thus, the ancients considered the far-reaching influence of time in general and used eight characters to represent it as an umbrella term.

Therefore, the birth date in Chinese traditional culture not only partially reveals the spatial and temporal environmental factors of the real world, but also shows the trend of the influence of these factors on people. Due to the probability and bias in the prediction and analysis, in most cases, it is only used as secondary reference in decision making. Nevertheless, when facing the traditional culture different from modern science, we should not simply reject or deny it. Instead, we should analyze the historical influence and essential characteristics behind it with scientific thinking skills.

Chapter 7
Gimbal-Based Attitude Determination

In Chap. 5, the methods of direct attitude determination using multiple-vector are introduced. If the vector used is obtained by observing the external reference, it will be limited by the observation conditions. To achieve accurate and stable attitude determination under various conditions, it is necessary to adopt an autonomous attitude determination method.

Gimbal-based attitude determination is the most important autonomous attitude determination method. The so-called gimbal platform refers to the frame platform using an inertial stabilization component to maintain stability. One direction of stability control can be achieved by one axis, and multi-axis stability control can be achieved by multiple axes. Through these stabilization axes, the reference coordinate system can be reproduced and controlled.

In some applications, such as aerial photography, gravity measurement and other fields, the stabilization platform also needs to be able to support some load. However, when it is applied to attitude determination, the load capacity requirement is not prominent. It is more important to obtain the high accuracy of tracking and repetition of the reference coordinate system in the platform coordinate system. The system can directly measure the angle relationship between the platform coordinate system and the body coordinate system through the angle measuring device and then obtain the attitude information.

The platform entities of such systems are flexible and diverse and do not necessarily have intuitive platform structures. Therefore, this kind of system is called a gimbal system. Because of the close relationship between the stabilized platform and the platform attitude determination, this chapter introduces them at the same time.

© Science Press 2024
H. Bian et al., *Essentials of Navigation*, https://doi.org/10.1007/978-981-99-5636-4_7

7.1 Inertial Stabilization Platform

The inertial stabilization platform or gyro stabilizer is a mechanical servo system that uses a gyroscope to control the stabilized platform to control the three axes of the platform to the preset coordinate system. Gyroscopes can be used to control stable platforms because of their unique characteristics.

7.1.1 Basic Characteristics of Gyroscopes

Generally, any device that can measure the rotation relative to inertia space can be called a gyroscope. Traditional rotor gyroscopes are defined as objects that rotate at high speed around the symmetrical axis of the body of revolution.

The gyroscope is composed of a high-speed rotating gyro rotor mounted on a suspension system so that the spindle of the gyro rotor has one or two degrees of freedom in space. Although there are many forms of the suspension system, the two-degree-of-freedom mechanical rotor gyroscope can be illustrated as the structure shown in Fig. 7.1. The suspension system consists of an inner ring, an outer ring and a base. The gyro rotor is supported by a spindle on the inner ring. The inner ring is supported by the inner ring axis on the outer ring. The outer ring is supported by the outer ring axis on the base. The rotor can rotate upward and downward around the inner ring axis or rotate left and right around the outer ring axis. Because it has two degrees of freedom, it is also known as a two-degree-of-freedom gyroscope (in Russia, it is called a three-degree-of-freedom gyroscope). This structure can make the spindle point in any direction in space. This support structure is called universal support or Cardan suspension.

For convenience of research, the spindle and inner ring axis of the gyroscope are defined as the x-axis and y-axis, respectively. Taking the center of the rotor as the origin, the $OXYZ$ of the gyro coordinate system is established. Only if the outer ring axis OZ_p is perpendicular to the OXY plane does the outer ring axis OZ_p coincide with the OZ. The three characteristics of the gyroscope are introduced below.

Fig. 7.1 Mechanical rotor gyroscope with two degrees of freedom. Photograph by Bian H. W.

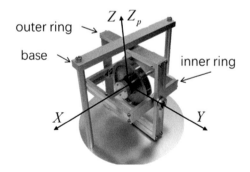

Gyroscopic Inertia

As the rotor rotates around its spindle at high speed, the inner ring axis of gyroscope *OX* will remain in the initial orientation in the inertial space. When the rotor rotates at high speed and an instantaneous impulse moment is applied to the gyroscope, the gyro spindle *OX* will produce a high frequency and micro oscillation near the original position and have only a slight deviation from the initial position. That is, once the gyro spindle is pointed in a certain direction, it will remain in the initial direction when there is no external torque on it. This feature can be used to establish a stable directional reference.

Precession

When the rotor rotates at high speed, if a constant torque (M_f) is applied along the inner ring axis of gyroscope *OX* (as shown in Fig. 7.2), the gyroscope is not intended to rotate around the *OY* axis but actually turns around the outer ring axis *OZ* perpendicular to the *OY* axis at an angular rate, that is, the spindle *OX* draws close to the external moment vector in the shortest way. The relation between external torque, angular momentum and angular velocity can be expressed by formula (7.1).

$$M_f = \omega_z \times H \qquad\qquad (7.1)$$

Generally, precession means that when a moment is applied to the gyroscope spindle, the gyroscope rotor does not rotate along the direction of the external force as people intuitively thought but turns in the direction of the shortest radial route to the external moment vector. Obviously, through precession, the turning rate and direction of the gyro can be precisely controlled by the external torque.

Fig. 7.2 Precession of gyroscope

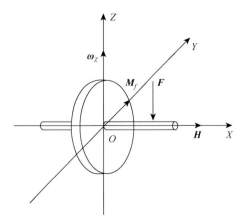

Gyroscope Reaction Moment

The third characteristic of the gyro is the gyroscope reaction moment (G.R.M). According to Newton's third law, when an object is moved by an external force, the object produces a reaction force with equal magnitude and opposite direction that applies to the object applying the external force. Similarly, when the gyroscope is processed by an external moment, the gyroscope itself also produces a reaction moment to the object by applying the external moment, which is called the gyroscope reaction moment. The G.R.M M_G is equal to the external moment M_E in magnitude but opposite in direction.

7.1.2   Types of Gyroscopic Stabilization Systems

There are many kinds of gyroscopic stabilization systems that can be divided into direct gyroscopic stabilization systems, dynamic gyroscopic stabilization systems and indirect gyroscopic stabilization systems.

Direct Gyro Stabilization System

The direct gyro stabilization system uses the G.R.M directly to balance the disturbance moment. Early direct gyroscopic stabilization systems were used to design ship stabilizers (such as the invention by Sherrick, a German, 1904) and aircraft stabilizers (such as the invention by Sperry, an American).

Taking a ship stabilizer as an example, its design idea is to install a single-degree-of-freedom gyroscope with a large momentum rotor (gyro) in the hull, and the frame axis of the gyroscope is consistent with the lateral axis of the hull. The gyro spindle is perpendicular to the deck plane. The hull can be regarded as a single-degree-of-freedom gyroscope "outer ring", and the longitudinal axis of the hull is the "outer ring axis", so the hull together with the gyroscope can be regarded as a huge two-degree-of-freedom gyroscope. When there are wave disturbance moments apply along the longitudinal axis of the hull, to make the hull roll, the gyroscope will rotate around the frame axis. The G.R.Ms generated by the precession are balanced with the wave disturbance moments so that the hull does not roll. Because the disturbance moment is totally compensated by the G.R.M, it is required that the gyroscope has a relatively large momentum moment, that is, the gyroscope has a large volume and weight. However, the direct gyroscope stabilization system can only overcome the disturbance moment, which changes alternately in direction. If the direction of some disturbance moments is constant, the gyroscope will always continue rotating in one direction. When the precession angle reaches 90°, the spin axis of the gyroscope coincides with the axis of the disturbance moment, and then the stabilization effect will be lost. Direct gyroscope stabilization systems can only be used in situations where the stabilized object is small and the precision requirement is low, and they are basically not used now.

Dynamic Gyro Stabilization System

The dynamic gyro stabilization system uses the gyroscope reaction moment and the external mechanical moment together to balance the disturbance moment. The dynamic gyroscope stabilization system is also an earlier gyroscope stabilization system. It has a stabilization loop. When there are disturbance moments, it relies on the gyroscope moment and the compensating moment produced by the motor in the stabilization loop to balance together. The gyroscope reaction torques play a major role in the initial stage, and the motor-generated moment is used to maintain stability. The stabilization loop of this system is relatively simple, and its accuracy is not very high.

Indirect Gyro Stabilization System

The indirect gyro stabilization system only uses the external mechanical moment to counteract the disturbance moment. With the rapid development of electronic, automation and computer technology, indirect gyro stabilization systems have become the mainstream technology of stabilization platforms. It is mainly composed of a high-precision gyroscope and a high-precision fast servo system, which can achieve higher accuracy. The volume of the gyroscope used in this system is usually very small, and the counteraction of the gyroscopic moment to the disturbance moment is negligible, but the system has a very fast reaction speed. When there is disturbance moment, the torque motor in the system can quickly generate a compensating moment to balance the disturbance moment.

7.1.3 Single-Axis Gyro Stabilization Platform

Principles

According to the number of stabilization axes of the stabilization platform, the stabilization system can be divided into single-axis, double-axes, three-axis and four-axes stabilization systems. A single-axis gyroscope stabilization system can make the stabilized object remain stable around one certain axis in space (Fig. 7.3). The single-axis gyroscope stabilized platform shown here is a dynamic gyroscope stabilizer. The dynamic gyroscope stabilizer uses the moment generated by the moment motor to balance the external disturbance moment applied on the platform (Fig. 7.4).

In Fig. 7.3, platform P is the platform needed to keep level. When the Y-axis has external force F to make P out of level, the disturbance moment M_f along the Z-axis

Fig. 7.3 Diagram of gyro
stabilization platform, with
permission from reference
[China Science Publishing &
Media Ltd. (CSPM)]

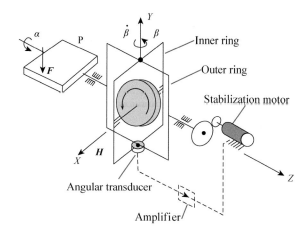

Fig. 7.4 Stabilization loop
of single axis dynamic gyro
stabilizer

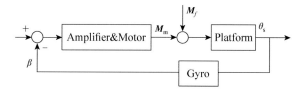

will be generated by F. According to the procession principle, the gyro spindle will produce precession angular velocity ω_Z along the Z-axis. Under the action of ω_Z, the gyro rotor will rotate with angle β. Formula (7.1) shows that M_f determines the magnitude and direction of ω_Z; ω_Z determines the magnitude and direction of turning angle β. Therefore, β can be used as a control signal to control the stable motor to produce the stable moment M_m used to balance the disturbance moment M_f and to maintain the platform keep level.

Characteristics

The dynamic gyroscope stabilizer can still achieve good performance when the technical requirements of the gyroscope and stabilization loop are not accurate enough. This is the reason why various dynamic gyroscope stabilizers have been used since the 1930s. The angular momentum of the gyroscope H can be smaller, the transition process of the system can be slightly longer, and the precession angle β *is* generally in the range of tens of minutes to several degrees when a dynamic gyroscope stabilizer is adopted. In some systems, especially in some small-scale systems, moment motors are not used, but torquers are used. In the gun stabilizer and other large devices, a hydraulic transmission device is used.

   The above dynamic gyroscope stabilizers can only achieve stability around the outer ring axis OZ, so it is a single-axis stabilization platform. Furthermore, by applying another moment to control the rotation of the gyro spindle, it is also possible to control the single-axis stabilization platform to be stable in any desired direction to satisfy different requirements, such as directional photography and weapon aiming.

On the basis of a single-axis stabilized platform, another gyroscope, stabilization loop and control loop can be added to establish another stable axis control to form a double-axis stabilized platform. By further increasing controlled axis, a three-axis gyroscope stabilization platform can be established that achieves three-dimensional stabilization and forms a three-dimensional reference. Therefore, single-axis stabilization systems are widely used in multi-axis stabilization platforms and inertial navigation systems.

Main Specifications of the Stabilization System

Torque rigidity and steady-state stiffness are key specifications of gyro stabilization systems. The ratio of the constant disturbance moment to the system steady-state error angle is called torque rigidity. The larger the torque rigidity is, the smaller steady-state error angle is when the same constant disturbing moment acts on the stabilized axis, and vice versa. Torque rigidity is similar to the stiffness of an elastic torsion stalk; the greater the stiffness is, the smaller the angle that can be twisted when twisting the torsion stalk with the same torque.

When the constant interference moment is applied to the stable axis of the system and the system reaches steady state, the ratio of the gyroscope precession angle to the constant disturbance moment causing error is called the steady-state stiffness. The steady-state stiffness indicates how much precession angle the gyroscope needs to produce the corresponding balance stable moment under the same magnitude of disturbance moment. Steady-state stiffness is the amplification factor from the precession angle of the gyroscope to the formation of a stable moment. To improve the steady-state stiffness of the system, it is necessary to increase the amplification factor.

7.2 Gyrohorizon and Gyrocompass

The two-axis gyroscope stabilization system mainly keeps the stabilized platform stable around two nonparallel axes in space. The two nonparallel stabilization axes can form a stable plane, so the two-axis gyroscope stabilization system is also called a two-axis stabilization platform. The two-axis gyroscope stabilization system is often used in ships and aircraft, such as gyrohorizon and ship level. The two-axis stabilization system can be used to maintain and indicate the vehicle attitude because it can indicate the local geographic horizon. The gyrocompass is also a two-axis stabilization system that can stably track and maintain the local meridian.

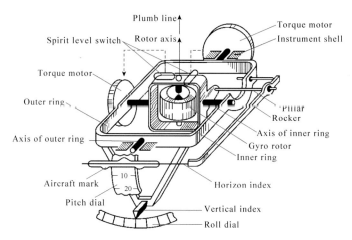

Fig. 7.5 Structure of gyro horizon, with permission from reference [China Science Publishing & Media Ltd. (CSPM)]

7.2.1   Gyrohorizon

The Composition of the Gyrohorizon

Gyrohorizon is a kind of flight instrument that uses a gyroscope to measure the pitch and roll of aircraft. It is necessary to establish a vertical or horizontal reference on the aircraft to obtain attitude [22]. By using the gyroscopic inertia, the gyro rotor axis can point to a fixed position in inertia space, which can be used as a directional reference. However, this directional reference obtained by a gyroscope cannot be used on Earth because of the rotation of the Earth in inertia space. Therefore, the gyroscope cannot automatically find and stabilize in the zenith line on Earth. At the same time, the friction moment applied on the inner and outer rings of the gyroscope will cause the gyroscope rotor axis to drift according to the precession of the gyroscope. Pendulum is used to sense the zenith line, but it will produce great errors when disturbed by acceleration. Therefore, inertial attitude determination equipment mostly combines a gyroscope and pendulum to solve the problems above and realize attitude determination. Figure 7.5 is the structure schematic diagram of the gyrohorizon on the aircraft. It consists of a two-degree-of-freedom gyroscope, pendulum sensing elements, torquers and indication dials.

Basic Working Principle

The outer ring axis of the gyro is installed parallel to the longitudinal axis of the aircraft. When the aircraft pitches or rolls, the instrument shell will rotate with the aircraft. At the same time, the gyro spindle still keeps its original orientation unchanged according to gyroscopic inertia; that is, the gyro spindle still indicates the vertical direction of the ground. Therefore, the attitude of the aircraft can be displayed intuitively and vividly using indication dials of the gyrohorizon, which uses the gyro spindle to indicate the zenith line. The liquid switch (in Fig. 7.5) mounted on the inner ring axis of the gyroscope is a bubble level sensor that acts as pendulum and works as a switch. The sealed container of the liquid switch is filled with a special conductive liquid and bubbles, and mutually insulated electrodes are installed at both ends of the container. When the disturbance moment is applied to change the direction of the gyro spindle (according to the procession), the two liquid switches sense the deviation angles of the gyro rotor axis relative to the zenith line and transform the angles into electrical signals. After amplification, the output signals are sent to the torquers installed on the inner and outer ring axes to generate a correction moment to balance the disturbance moment so that the gyro spindle always points to the zenith line. The correction system adopts the method of intersection correction: when the spin axis is deflected relative to the zenith line around the inner ring axis, the correction moment is applied to the outer ring axis, and vice versa.

Because the correction speed is slow, generally less than $10°/m$, when the liquid level of the liquid switch is inclined due to the acceleration disturbance of the aircraft, the error correction in a short time only causes the rotation axis to deviate from the zenith line at a very small angle. Moreover, when the linear acceleration or hovering angular velocity of the aircraft exceeds a certain value, the correction circuit will be automatically cut off to prevent a larger correction error and improve the anti-interference capability. The gyro spindle is in a random position before the instrument starts. To make the spindle reproduce the zenith line quickly, the correction moment can be increased or the spindle can be locked in the vertical direction by a locking device.

To prevent the pitch angle from being $90°$ where the outer ring axis coincides with the rotation axis, the gyroscope loses its normal working condition. A servo ring is added to the aircraft horizon meter. The gyroscope rotor and inner and outer rings are installed on the servo ring, while the servo ring axis is installed parallel to the longitudinal axis of the aircraft. The servo ring can guarantee the orthogonality of the automatic axis, the inner ring axis and the outer ring axis in any attitude maneuvering aircraft so that the demonstration range of pitch and roll can reach $360°$.

The gyroscopic horizon can be divided into the direct reading type and remote reading type. The airplane attitude is displayed directly on the indication dials in the direct reading type. The remote reading type outputs the aircraft attitude signal through the sensor mounted on the gyroscope and transmits it to the horizon indicator by the remote transmission system for display. This kind of gyroscope with signal sensors is called a vertical gyroscope. As an attitude sensor, it can provide pitch and roll signals to airborne systems. Planes use the direct-reading horizon. When the plane climbs, the aircraft sign moves below the horizon. When it dives, it does not conform to the intuitive sense. The remote-reading horizon can overcome this shortcoming.

7.2.2 Gyrocompass

Basic Information

With the development of navigation and shipbuilding technology, the emergence of steel for the construction of ships, especially large and medium-sized ships, makes the reliability and accuracy of magnetic compass far from meeting the requirements and makes the urgent desire to pursue higher precision and more stable heading-indicating equipment. This led to the advent of a new heading-indicating instrument, gyrocompass. In 1908, the world's first gyrocompass was manufactured by Anschuetz, Germany. In 1909, Sperry developed the second practical gyrocompass in the world. Gyroscope heading-indicating technology can be said to be a milestone of the whole human heading-indicating technology; it is an epoch-making invention. It significantly improves the human heading-indicating ability and has become necessary navigational equipment for ships [23].

A gyrocompass is a kind of equipment that uses gyroscopic inertia to measure the earth rotation vector, uses a pendulum sensor to sensitize the local gravity vector, and provides a northward reference by employing a control mechanism and damping mechanism. Similar to the celestial multi-vector attitude determination system, gyrocompass utilizes two kinds of vectors, the earth rotation vector and the gravity vector. However, unlike the celestial multi-vector attitude determination system, the gyrocompass directly establishes a northward reference and indicates heading by controlling the rotation of the gyro spindle. This section introduces the north-finding principle of the gyrocompass, which is the key principle of various inertial heading and attitude systems.

North-Finding Principle of Gyrocompass

Gyro Apparent Motion

The angular velocity of the Earth's rotation ω_e affects the movement of the local geographical coordinate system. For ease of analysis, ω_e can be projected to the local geographical coordinate system and decomposed into vertical component ω_1 and horizontal component ω_2. In the Northern Hemisphere, the horizontal component ω_1 points to the north, also known as the northward component (Fig. 7.6). ω_1 along the ON direction, ω_2 along the OT direction (Eq. 7.2).

$$\begin{cases} |\omega_1| = |\omega_e| \cos \varphi \\ |\omega_2| = |\omega_e| \sin \varphi \end{cases} \tag{7.2}$$

The meaning of ω_1 in the Northern Hemisphere is the rotation of the horizon plane ONW along the ON axis from left to right. ω_2 is the rotation of the meridian

Fig. 7.6 Decomposition of
the angular velocity of the
earth's rotation, with
permission from reference
[China Science Publishing &
Media Ltd. (CSPM)]

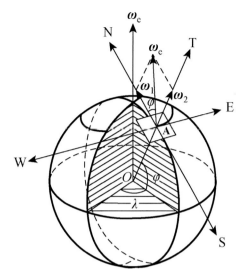

plane ONT along the zenith axis (OT axis) in ω_2 from the east and west. Because of gyroscope inertia, the gyro spindle will keep the fixed direction relative to inertial space without external torque. Therefore, the earth will have relative motion relative to the gyro spindle. As people stand on earth, the motion of the gyro spindle relative to the earth is observed, which is called apparent motion.

When the projection of the gyro spindle on the horizon plane is located to the east of the ON axis, because the horizon plane rotates from east to west at angular velocity ω_1 around the ON axis and the gyro spindle points invariant because of gyroscope inertia, the spindle vertical angle will gradually become larger visually. Similarly, when the projection of the gyro spindle on the ground plane is located to the west of the ON axis, the spindle height angle will gradually decrease. If the projection of the gyro spindle on the ground plane is on the ON axis, the angular velocity will not affect the vertical angle of the gyro spindle because the ON axis is the rotation axis of the ground plane, and the elevation and reduction of the spindle will no longer occur in the apparent motion. In other words, only in the position where the spindle is not in the true north-east or west, will the spindle appear elevated or decreased. This unique gyro feature can help us to obtain a solution to the problem of finding the north.

Principle of Gyro North-Finding

It is difficult to directly determine whether the spindle deviates from the north. However, by utilizing the feature that the gyro spindle will inevitably rise or decrease when it does not point to the north, the difficult problem of finding the north can be changed into a relatively easy problem of judging whether the elevation angle of

Fig. 7.7 Motion trajectory
of spindle without damping

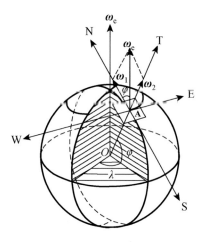

the spindle is larger or smaller. The following control schemes can be adopted to
address the problem: (1) when the spindle points to the east, the spindle is raised
and gradually higher than the horizon, and a moment can be applied to control the
gyroscope moving westward; (2) when the spindle points to the west, the spindle
decreases and gradually falls below the horizon, and a moment can be applied to
control the gyroscope moving eastward; and (3) when the spindle is in the meridian,
the elevation angle of the spindle remains unchanged, and no procession control
moment is needed. The procession control moment produced in the above control
schemes is called the north-seeking moment.

As long as the north-seeking moment is proportional to the elevation angle θ,
north-seeking can be realized. A pendulum can be used to determine the elevation
angle θ. Once the spindle is found to be raised, a westward driving moment is
applied to control the gyro spindle to move northward. If the spindle is lowered,
an eastward driving moment is applied to control the gyro to move back northward.
When the spindle is raised upward, the spindle moves westward, and when the spindle
is lowered downward, the spindle moves eastward. After a complete cycle, the spindle
motion will depict the trajectory of an ellipse centered on the north-south axis in the
horizontal plane (Fig. 7.7). This is the basic north seeking idea of gyrocompass. In
Fig. 7.7, v_1 is the apparent motion caused by ω_1, v_2 is the apparent motion caused by
ω_2, and v_3 is the spindle procession caused by the north-seeking moment. According
to the principle of proportional control, when the gyroscope spindle deviates from
the north, it will oscillate around the north, as shown in Fig. 7.7.

Figure 7.7 show that the gyro spindle cannot point to the north by employing
only the north-seeking moment. We need another moment to make the gyro spindle
hold in the meridian. This moment is the damped moment. The damped moment
usually adopts the azimuth moment, which is proportional to the position of the
spindle elevation angle θ. Damping indirectly reduces the amplitude of precession.
The typical damping time constant is 60~90 min. Due to the influence of gravity and

the damped moment, the gyro spindle can gradually stabilize to the direction of true north, thus achieving the purpose of finding north.

According to the above principle, the gyrocompass can achieve self-alignment, and the heading error will not increase with time. The typical gyrocompass stable time is approximately 1 h.

Types and Characteristics of Gyrocompass

Types of Gyrocompass

According to the different ways of applying the controlled moment, gyroscopes can be divided into two categories: gravity moment and electromagnetic moment, which are called pendulum gyrocompass and electromagnet control gyrocompass, respectively.

(1) Pendulum gyrocompass

The mass center of the gyro ball, which is in the gyroscope in the gyrocompass, is below its geometric center to produce a gravity moment applied to the horizontal axis of the gyrocompass. This moment is used to realize automatic north finding and stable tracking. A pendulum gyrocompass designed by Anschuetz is shown in Fig. 7.8. Its control process is as follows: when the spindle is raised, the spindle is on the east side of the meridian, the mass center is below the horizon, and the gravity force applied in the gravity center (point G in Fig. 7.8) produces a pendulum moment because there is a distance l between the mass center and geometric center. Through the right-hand rule, the direction of the pendulum moment is pointed to the west and makes the spindle rotate to the west according to the precession principle. When the spindle is moving westward, the spindle gradually drops below the horizon, and the pendulum generates the pendulum moment to control the spindle moving eastward. The pendulum gyroscope has the ability to find the north. If we want the spindle to point north stably and track the true north rotation of the earth, the gyroscope spindle needs to maintain a small elevation angle to generate westward momentum, which is exactly equal to the angular velocity of the gyroscope spindle rotating along the true north of the earth. This state is stable and can be used to track the north stably.

The design of the Anschuetz gyrocompass is exquisite and has wide influence. Gyrocompass has high directional accuracy and strong load capacity, but the whole set of equipment is relatively large. Since the advent of the gyrocompass, there have been nearly 100 models. There are three main series: Anschuetz series, Sperry series and Arma-Brown series. In addition to the three above major gyrocompass series, there are DTG gyro compasses (such as SKR-80 in Norway and cruise-type in Ukraine, etc.) and magnetic suspended gyro compasses (such as pirate-type in Russia).

Fig. 7.8 Anschuetz
pendulum-style gyrocompass

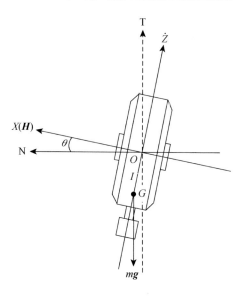

(2) Electromagnet control gyrocompass

Electromagnet control gyrocompass developed in the late 1950s. Unlike the
pendulum gyrocompass, the electromagnet control gyrocompass achieves automatic north finding and stable tracking by using torquers to generate appropriate
electromagnetic control moments.

Unlike the pendulum gyrocompass, the circuit parameters can be changed when
the electromagnet control gyrocompass is started, and the horizontal north-seeking
moment and damped moment coefficient of the gyrocompass can be adjusted to
shorten the stable time of the gyrocompass. During the manufacture of the electromagnet control gyrocompass, improvement measures are taken from the structure and circuit. Generally, the electromagnet control gyrocompass has two working
states, that is, "azimuth finder" and "compass". At high latitudes, the working state
of "azimuth finder" must be used. Therefore, this kind of gyrocompass is also called
the bimodal compass.

Error of Gyrocompass

(1) Latitude error

Earth rotation will cause the gyro spindle to deviate from the north–south direction at a
rate proportional to the sinusoidal latitude, and we adopt north-seeking and damping
control loops to offset this deviation. However, this correction still lags behind,
resulting in latitude-related heading deviation of the compass. This latitude deviation
can be eliminated by applying latitude-related compensation torques. Therefore, the
gyrocompass needs latitude information support.

(2) Speed error

A ship navigates on the surface, and the x-y plane of its body coordinate system always remains parallel to the local sea level, which is equivalent to the slow rotation relative to the inertial space around its pitch axis. When the ship is sailing along the north–south direction, the gyro spindle of the gyrocompass points north, and the gyro rotor pitch axis is consistent with the ship pitch axis and does not rotate with the ship. As a result, the pitch axis around the ship is slowly rotated relative to the hull, and the gravity measurement component of the gyrocompass will measure this increased horizontal deviation, controlling the gyro to produce a heading disturbance proportional to $v_{eb,N}^n / \cos \varphi$ A north-south velocity of 20 m/s will be introduced with an approximately $1°$ deviation in the mid-latitude region.

This error due to the introduction of the northward velocity can be eliminated by correcting the gyrocompass output or by increasing the control torque to control the gyro spindle realignment. Therefore, the gyrocompass requires input of the northward velocity information. When the ship is sailing along the east-west direction, the pitch axis of the ship is consistent with the spindle of the compass gyro rotor, so the rotation of the local horizontal plane of the ship will not affect the control loop of the gyrocompass. Because the speed of the aircraft is fast, the speed error cannot be effectively compensated, so the gyrocompass is not suitable for the aircraft. The speed error also prevents the gyrocompass from working at the earth poles.

Characteristics of Gyrocompass

Compared with the magnetic compass, the gyrocompass has many advantages, such as high accuracy, stability and reliability, can connect multiple repeaters, can work continuously for a long time without correction, is helpful to ship automation and is not affected by magnetic interference. The gyrocompass directly provides the true north heading reference, and its heading data can be used by several systems, such as radar systems, sonar systems, control systems and weapon systems. However, there are also some shortcomings, such as a complex structure, long start-up time, and poor accuracy at high latitudes.

7.3 Stabilized Gyrocompass

There are many inertial equipment that can measure heading and attitude, such as the attitude and heading reference system (AHRS) or stabilized gyrocompass. A stabilized gyrocompass is a kind of ship navigation equipment that provides heading and attitude information. Based on the gyrocompass, an inertial stabilization platform is designed to track and indicate the geographic coordinate system and to measure the attitude. It not only provides reliable heading information for ship navigation but also provides heading and attitude for radar and weapon-launching.

With the development of technology, the strapdown attitude system has been widely used, and this system will be introduced in the strapdown inertial navigation section in Chap. 8. Here is a brief introduction to the stabilized gyrocompass.

The first stabilized gyrocompass was developed by the US company Sperry in 1949. Since then, many countries have spent considerable manpower and material resources innovating and have made a variety of models, such as the MK29, AM/ WSN-2 stabilized gyrocompass in the United States, the French MCV-3 stabilized gyrocompass and the German PL-41/MK3 stabilized gyrocompass. Compared with the magnetic compass and gyrocompass, the stabilized gyrocompass can provide high attitude accuracy, and the performance of inertial sensors is less demanding than that of inertial navigation systems.

7.3.1 Three-Axis Stabilization Platform

The stabilized gyrocompass adopts a three-axis stable platform to realize body attitude determination. The three axes of the stable platform are composed of the platform azimuth axis, the inner ring axis and the outer ring axis. According to the coordinate system simulated by the stability platform, the three-axis stable platform can be divided into a spatially stable platform and a tracking stable platform. The former realizes the simulation of the inertial coordinate system, which is not affected by the carrier motion and interference applied; the latter realizes the simulation of any navigation coordinate system needed, most of which is the local geographical coordinate system. To control the platform coordinate system to follow the local geographic coordinate system, the latter must apply a correction current to the gyroscope torquer. The stabilized gyrocompass mainly adopts the three-axis stable platform, which simulates the local geographical coordinate system.

The establishment of a three-axis stable platform requires 3 single-degree-of-freedom gyroscopes or 2 two-degree-of-freedom gyroscopes together with an electromagnetic pendulum or accelerometer as a gravity sensor. The early stabilized gyrocompass generally used an electromagnetic pendulum as a gravity-sensitive element. Electromagnetic pendulums have many advantages, such as a simple structure, small volume, light mass and low cost, but their performance is difficult to improve due to structural constraints. The appearance of a high-precision accelerometer overcomes the shortcomings of the electromagnetic pendulum and gradually replaces it.

Each two-degree-of-freedom gyroscope is capable of sensing the angular motion of the platform rotation in two directions, and the two sensitive directions are perpendicular to each other. The stabilized gyrocompass only uses the gyroscope as the sensitive element of the interference moment and no longer uses the gyroscope moment to compensate for the external interference moment directly. Therefore, a framework gyroscope with a large angular moment and large volume is no longer used, and the volume and mass of the gyroscope platform can be reduced as a result. Two two-degree-of-freedom gyroscopes can sense the interference torque in three directions of the platform perpendicular to each other and cancel the interference torque by the control torque generated by the corresponding stable motor. The redundant fourth sensitive axis can be used for correction or monitoring.

Figure 7.9 shows the basic composition of the simplified stabilized gyrocompass. The triaxial stable platform with gyroscopes and accelerometers installed consists of the outer ring (roll ring), the middle ring (roll ring) and the inner ring (azimuth ring). Two mutually perpendicular east and north accelerometers and two biaxial gyroscopes are installed on the stable platform. The current gyro platform mass ranges from dozens of kg to only 8 kg, and the outer dimension ranges from more than 0.5 m to only 0.08 m. The gyro-stabilized platform develops toward high precision, high reliability, low cost, and can compensate for the platform error.

On the carrier with large maneuver, the three-axis stable system has the phenomenon of gimbal locking. Figure 7.10(a) shows a triaxial system mounted on a missile. Its outer ring axis is consistent with the pitch axis direction of the missile. This platform allows the projectile to rotate ± 360° around the azimuth axis and pitch axis but only less than ± 90° around the roll axis. When the rolling angle is 90°, as shown in Fig. 7.10(b), the projectile will drive the outer ring to rotate 90°, so that the outer ring and the middle ring are in the same plane. At this time, the inertial

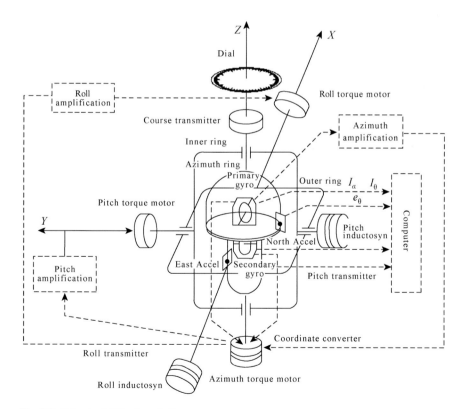

Fig. 7.9 Tri-axis stabilization platform

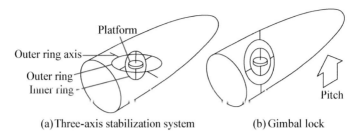

(a) Three-axis stabilization system   (b) Gimbal lock

Fig. 7.10 Gimbal lock of gimballed system

Fig. 7.11 Four-axis
stabilized platform

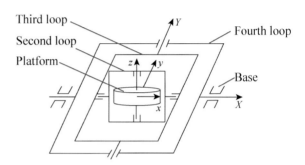

navigation platform loses one degree of freedom with only two degrees of freedom
left. The gimballed system cannot isolate the rotation of the projectile perpendicular
to the plane of the balance ring, which is called the phenomenon of the gimbal lock.
To avoid the possible gimbal lock of the three-axis stable platform, the four-axis
stable platform system must be selected (Fig. 7.11).

7.3.2 Principle of Stabilized Gyrocompass

The stabilized gyrocompass controls the three-axis stabilized platform to track the
geographic coordinate system by the gyroscope. When the three-axis stabilized platform stably points to the geographic coordinate system, the attitude angle of the
carrier can be obtained by measuring the rotation angle of the stabilized platform
relative to the body coordinate system through an angular transducer.

Working Principle of Control Loops

The north accelerometer A_N, the east gyro G_E and the azimuth gyro G_Z constitute
the compass loop to track the north level of the local geographic coordinate system
(seen in Fig. 7.9). The east accelerometer A_E and the north gyro G_N constitute the

eastern horizontal loop to track the eastern level of the local geographical coordinate system.

The stabilized gyrocompass has high horizontal accuracy. In the static base or wharf mooring state, when the three-axis stable platform deviates from the level, the gravity sensors installed on the platform can measure the gravity components in different directions. The measured value of the gravity component is used as the horizontal correction error signal, which is amplified by the amplifier and sent to the torquer of the gyroscope on the platform to control the gyroscope procession. Then, the stable platform is driven to track the motion of the gyroscope coordinate system through the platform stability loop so that the stable platform is always at the local level.

When using the accelerometer as a gravity sensor, because the accelerometer measurement is relative to the inertial space, the harmful acceleration and the moving acceleration should be extracted from the output signal of the accelerometer according to formula (7.3); then, the signal reflecting the horizontal deviation angle of the platform can be obtained.

$$\begin{cases} A_N = \dot{V}_N + W_{BN} + g_a \\ A_E = \dot{V}_N + W_{BE} - g_\beta \end{cases} \tag{7.3}$$

where:

A_N, A_E	North and East accelerometer measurements;
$\dot{V}_N, \dot{V}_E$	Northward and eastward acceleration of ships relative to the geographic coordinate system;
W_{BN}, W_{BE}	Harmful acceleration;
g_a, g_β	Horizontal component of gravity.

Using the accelerometer output that reflects the horizontal deviation angle of the three-axis inertial platform, proper control signals can be designed to control the east gyro, the north gyro and the azimuth gyro to procession so that the stabilized gyrocompass can track the local geographical coordinate system. The block diagram of the stabilized gyrocompass adopting this control principle is shown in Fig. 7.12.

The classical initial alignment process of a stabilized gyrocompass is usually divided into two steps: first, horizontal leveling, then azimuth alignment. The azimuth alignment is carried out on the basis of horizontal leveling, and the compass effect is generally adopted to achieve compass azimuth alignment. When the platform stably tracks the local geographical coordinate system, if the platform has a heading deviation angle along the azimuth axis, due to the compass effect, the platform will produce a tilt around the eastward axis, which can be measured by the north accelerometer. Using the output of the north accelerometer, proper control can be designed to control the rotation of the platform around the azimuth axis in the direction of reducing the azimuth deviation so that the platform can automatically find the north and complete the azimuth alignment.

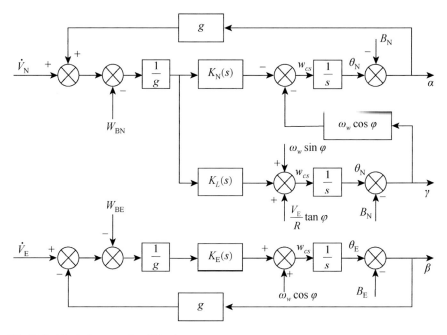

Fig. 7.12 Block diagram of stabilized gyrocompass

Principle of Stabilization Loops

Through the control loop and the stabilization loop, the stabilized gyrocompass has the function of indicating true north and level. The triaxial stabilization platform needs to send out the correct heading, pitch and roll through the coordinate converter. The coordinate converter is the main difference between the three-axis stable platform and the two-axis stable platform.

To explain the principle of the stabilization loop, it is assumed that the carrier is changed from north to east, and the azimuth sensor has signal output because the gyro spindle keeps pointing to north. This signal is processed by the azimuth stabilization loop, which generates a control voltage sent to the azimuth torque motor. Under the azimuth motor torque, the platform rotates around the azimuth axis, and the output signal of the azimuth sensor of the gyroscope is restored to zero. Therefore, despite the carrier changing course, the platform frame still dynamically stably points north (Fig. 7.13).

When the carrier is swaying, the platform frame remains level through the pitch and roll stabilizer loop and torque motor. The angle sensor of the north gyro and the east gyro can measure the angle between the gyro spindle and the platform along the north-south direction and the east-west direction. The pitch and roll motors are installed on the pitch and roll axes of the platform, respectively. Only when the ship is sailing along the north are the pitch and roll axes of the platform (body coordinate system) consistent with the north-south direction and the east-west direction

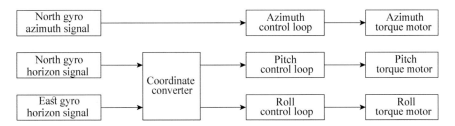

Fig. 7.13 Block diagram of stabilized gyrocompass stable loop

(geographical coordinate system). When the ship is sailing along another course, the horizontal deviation angle in the north-south or east-west direction must be converted into the deviation signal along the pitch and roll direction by means of a coordinate converter to realize the correct control of the attitude stabilization loop. The coordinate converter is implemented by a sinusoidal rotary transformer mounted on the azimuth axis.

Appendix 7.1: Questions

1. Please briefly describe the differences between gimbal-based attitude determination and vector-based attitude determination.
2. Please briefly describe the three characteristics of gyro and its application in gyro stabilized system.
3. Please briefly describe the different characteristics and applications of direct, dynamic and indirect gyro stabilized systems.
4. Briefly describe the physical phenomena and principles of the Foucault pendulum. In the southern hemisphere of the Earth, what is the vertical component of the Earth's rotation applied to the gyro's apparent motion? What is the connection with the Foucault pendulum?
5. Take the gyro azimuth and level instrument as an example to analyze the gyro azimuth-keeping principle and application.
6. Explain what is the gyrocompass effect? Why is the gyrocompass effect related to latitude? Please briefly describe the north-finding principle and working process of the gyrocompass.
7. Please briefly describe the composition and working principle of the stabilized gyrocompass.
8. By comparing the coordinate transformation of the navigation system learned before, please briefly describe the coordinate transformation of the control signal in the stability control of the horizontal loop.

Appendix 7.2: The Misunderstood World—"Seeing is Not Necessarily Believing"

The gyroscopic apparent motion indicates that people's feelings and intuitions are often vague and inaccurate. In fact, to understand the seemingly difficult things involves overriding the habitual feelings and intuitions. Having this type of control is indeed a breakthrough and upgrade in terms of thinking.

The history of human development has witnessed many examples where scientific rationality violates perceptual intuitions. Galileo's free-fall test, for example, proved Aristotle's intuitive analysis wrong. Copernicus' heliocentric theory indicates that the geocentric theory is an incorrect understanding of the world. With many similar cases, people have started to be aware of their own cognitive limitations, realizing that seeing is not necessarily believing. Therefore, scientific instruments are designed to enhance their cognitive ability, including microscopes, telescopes, infrared detectors, sonar, radar, and many other measuring devices. Scientific measurement has thus become the basis for human to understand the world and to research as well.

As a matter of fact, when sailing, people are more likely to make right decisions based on the accurate external information obtained with different measuring tools. Therefore, a good command of scientific measuring instruments and methods is crucial to students majoring in navigation, and so is the clear understanding of the value of accurate data and scientific ways. With the applications in many other university courses, this way of thinking will help students develop a structure of empirical knowledge and foster rich scientific literacy.

Chapter 8
Inertial Navigation

There is a story in "The Adventures of Sherlock Holmes": Holmes is kidnapped and blindfolded, tied up in a carriage. Holmes estimates the time spent by silently counting the time, and then calculates the distance based on the speed of the carriage to determine the approximate location of the kidnapper. The method used by Sherlock Holmes is Dead Reckoning (DR).

There is a physiological organ in human body that adopts a similar principle of DR. The vestibular organ of the inner ear is the balance receptor organ in human body, which can sense the changes of horizontal or vertical linear acceleration. When our vehicle rotates or turns, such as when the car turns and the aircraft makes circular motion, the angular acceleration acts on the corresponding semicircular canals on both sides of the inner ear. When hair cell in the ampulla of one side of the semicircular canals are stimulated to bend and deform to produce a positive potential, the opposite hair cell will bend and deform to produce a negative potential. Similarly, when the linear acceleration (deceleration) changes of the riding tools, such as car starting, deceleration braking, ship shaking, bumping, elevator and aircraft ascending or descending, these stimuli cause the deformation discharge of the hair cell in the capsule spots of the vestibular elliptical bursa and saccule to transmit and sense the operation information to the central nervous system. So, the eyes help us directly locate and navigate based on external information, while ears help us determine attitude and infer orientation.

Dead reckoning (DR) is the primary form of this approach, and inertial navigation is the advanced form.

In navigation technology, dead reckoning is particularly important as it does not require external information to realize navigation. Especially in many conditions where external information cannot be obtained, such as submarine navigation, night driving or external information is often disturbed, direct positioning methods cannot be used any more, but the navigation based on DR has become the only autonomous method that can work normally. It calculates the current or future position from past known positions based on the measurement of parameters such as the carrier's

© Science Press 2024
H. Bian et al., *Essentials of Navigation*, https://doi.org/10.1007/978-981-99-5636-4_8

heading, speed, and time, thereby obtaining a motion trajectory. Early days of sailing, the heading was determined by magnetic compass or compass, the velocity was determined by log, time was displayed in the chronometer, and the ship position was calculated by using these parameters. The INS widely used in modern times uses the accelerometer to measure the acceleration of the carrier, and by integrating the acceleration twice, the calculated position, velocity and attitude of the carrier can be continuously output.

8.1 Principle of Dead Reckoning

8.1.1 Brief Introduction of Dead Reckoning

Dead reckoning obtains the current position by measuring position changes or measuring velocity and integrating it with the original position (Fig. 8.1). Only by projecting the velocity and displacement on the carrier coordinate system to the reference coordinate through the attitude parameters can the velocity or range of the carrier relative to the reference coordinate system be calculated. For 2D navigation application, heading is sufficient. For 3D navigation application, it is necessary to measure three components of attitude parameters. When the attitude changes, the smaller the step size for position calculation adopt, the more accurate navigation parameters can be obtained. In the past, it was mostly done manually, which severely limited the data update rate. Now, it is done by computer.

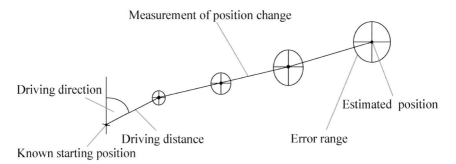

Fig. 8.1 Principle of dead reckoning

8.1.2 Principle of Dead Reckoning

Dead Reckoning of Ship

Dead reckoning is the basic method to obtain the position of the ship at any time in the voyage [24]. Under the condition of not relying on external navigation targets, relying solely on the most basic heading and speed indication equipment of the carrier, as well as external wind and current data, starting from the known starting point of the calculation, a certain precision speed and carrier position at a certain time can be calculated. Through trajectory calculation, the pilot can clearly understand the continuous trajectory of the ship's motion at sea. In modern ships, the function of manually plotting and automatically inferring the flight path is completed by a flight path plotter (see Chap. 10) or a navigation workstation.

(1) Classification

There are two kinds of dead reckoning. The first is track plotting, which is used to solve the two problems. One is plotting the ship trajectory according to heading, speed, sailing distance, wind and current. The other is plotting the ship heading and obtaining the estimated position according to the planned route, wind and current. This method is simple and intuitive, and it is the main method for the pilot to calculate ship position during sailing.

The second is track calculating, which uses mathematical equations to calculate the ship position and then plots the ship position on the chart according to heading, speed, sailing distance, wind and current. This method is the theoretical basis of computer track calculation and ship steering automation.

(2) Basic method

Given the compass heading, log range, and wind and current elements, calculate the trajectory direction. The effects of wind and current on ship navigation often coexist. The comprehensive impact of wind and current on ship navigation is discussed based on a separate analysis of the effects of wind and current. Due to the different characteristics of the influence of wind and flow on ship navigation, when analyzing the combined effects of wind and flow, the influence of wind is generally analyzed first, and then the influence of flow, that is, the drawing method of "wind before flow" is adopted.

As shown in Fig. 8.2, when the ship operates with speed V_E and heading C_T, under the influence of the wind, the ship deviates from a leeway angle α and moves along the route line in the wind at speed V_α. As a result of the current, the ship deviates from a drift angle β. A drift triangle is made on the route line in the wind to obtain the final velocity V_γ and track in the wind and current.

The true course and speed should be replaced by the track in wind and speed in wind when the wind and current simultaneously affect the drift triangle. The six elements of the drift triangle are track in wind, speed in wind, current direction, current velocity, track in wind and current, and speed in wind and current. Speed V_E is usually used instead of wind speed V_α when the wind speed is unknown.

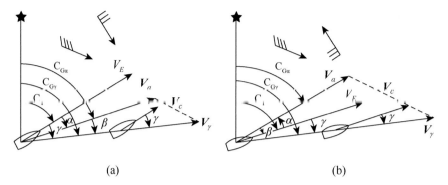

Fig. 8.2 Wind and current pressure difference

To calculate a ship's position by sailing in the wind or current, the range can be calculated based on the range S_L or wind speed recorded ($V_\alpha \times \Delta t$) by the log. A point can be measured on the wind track line, and then a flow parallel line intersects with the wind track line from that point. The intersection point is the calculated ship's position at that time. The range in the wind can also be calculated based on the obtained speed and sailing time in the wind, and the ship position can be directly calculated by intercepting it on the track line in the wind.

Finally, it should be noted that under the combined influence of wind and current, although the ship moves along the track line in the wind and current, the direction indicated by the bow is still the direction of the original heading line.

Dead Reckoning of Pedestrian

DR of pedestrian is one of the most challenging applications in navigation technology. The pedestrian navigation system must be able to operate normally in urban areas, under trees, and even indoors where GNSS and most other radio navigation systems have poor performance. Inertial sensors can be used to measure forward motion through dead reckoning. However, pedestrian applications generally require small size, light weight, low power consumption, and in most cases, low cost, therefore MEMS sensors are often used. When used MEMS alone, the performance of MEMS is poor, resulting in limited accuracy of calculation. Low dynamic and high vibration environmental conditions also limit the effectiveness of GNSS or other positioning systems in correcting effect.

8.1.3   Characteristics of Dead Reckoning

Because dead reckoning requires the use of previously known positions, heading, and estimated speed to derive the current position over time, the new position calculated is only derived from the previous step value, and the error and uncertainty of the derived position estimation value increase with time. Therefore, it is often used in short time or low precision situations.

8.2 Principle of Inertial Navigation

Inertial navigation uses inertial sensors (gyroscopes, accelerometers) to measure the linear and angular motion of the carrier relative to the inertial space. When the initial conditions are given, the output of the carrier's heading, attitude, speed, position and other navigation information can be provided. INS (INS) is a powerful navigation equipment with the advantages of autonomy, concealment and so on. It is the most important mean of navigation and positioning of underwater vehicles and plays an irreplaceable role in military applications.

8.2.1 Basic Principles of Inertial Navigation

Citation

Let the carrier move uniformly in a straight line, the motion time be t, the velocity be V, the acceleration be a and the travel distance be S, the relationship between the above parameters is as follows:

$$V = V_0 + at$$

$$S = S_0 + V_0 t + \frac{1}{2} a t^2 \qquad (8.1)$$

When all the initial condition is zero, i.e., $S_0 = V_0 = 0$ (S_0 is the initial distance and V_0 is the initial velocity), then:

$$V = at$$

$$S = \frac{1}{2} a t^2 \qquad (8.2)$$

Because the carrier's heading and speed change at time, the acceleration a will not be a constant and can no longer be solved by algebraic operation at this time but need to continuously measure the acceleration a and perform recursive calculation. The recursive calculation process is as follows:

(1) **The acceleration is integrated once to obtain the velocity of the carrier (velocity update):**

$$\begin{cases} V_E(k+1) = V_E(k) + \int_0^\tau a_E(k)dt \\ \\ V_N(k+1) = V_N(k) + \int_0^\tau a_N(k)dt \end{cases} \tag{8.3}$$

where $V_E(k+1)$ and $V_N(k+1)$ are the *No. k* + 1 recursion values of the eastward and northward velocities of the vehicle. $V_E(k)$ and $V_N(k)$ are the *No. k* recursion values of the eastward and northward velocities of the vehicle. $a_E(k)$ and $a_N(k)$ are the *No.* k recursion values of the eastward and northward accelerations of the vehicle. τ is the sampling time interval.

(2) **The velocity is integrated to obtain the range of the carrier (position update):**

$$\begin{cases} \varphi(k+1) = \varphi(k) + \frac{1}{R} \int_0^\tau V_N(k)dt \\ \\ \lambda(k+1) = \lambda(k) + \frac{1}{R\cos[\varphi(k)]} \int_0^\tau V_E(k)dt \end{cases} \tag{8.4}$$

where $\varphi(k+1)$ and $\lambda(k+1)$ are the *No. k* + 1 recurrence values of the latitude and longitude of the vehicle. $\varphi(k)$ and $\lambda(k)$ are the *No. k* recurrence values of the latitude and longitude of the vehicle.

This is the basic principle of the horizontal north pointing semi-analytic INS.

Platform Coordinate System and Attitude Determination

The difficulty to realize inertial navigation is how to keep the measuring direction of the accelerometer always along the east and the north under the condition of carrier maneuver, although the basic principle of inertial navigation is very simple. The key to solving this problem is to build the three-axis inertial stabilization platform. When the platform tracks the local geographical coordinate system stably, the three axes of the platform always point to the east, north and zenith. In this way, the carrier attitude can be obtained and output directly by the angle measuring elements installed on the gimbal of the platform, and the sensitive axis of the accelerometer A_E and A_N can always point to the east–west or north–south direction respectively even in maneuver process. Once the platform coordinate system cannot be guaranteed to be consistent with the geographical coordinate system, it will bring errors to the accelerometer measurement and the carrier attitude angle measurement. So it is very important to establish an accurate platform coordinate system.

How to ensure platform level and north pointing, that is, how to keep stable platform three-axis tracking geographic coordinate system. This issue is described in Chap. 7. It is assumed that the stable platform has achieved the alignment of the geographical coordinate system according to the working mode of the stabilized gyrocompass intróduced in Chap. 7. Taking the moving base as an example, when the carrier P moves on the Earth surface (Fig. 8.3), the rotation angular velocity of the geographical coordinate system ENT with P as the origin relative to the inertial space will consist of two parts: one is the rotation angular velocity of the Earth $\boldsymbol{\omega}_{ie}$, that is, the rotation angular velocity of the Earth relative to the inertial space; the other is the rotation angular velocity $\boldsymbol{\omega}_{en}$ caused by the movement of the carrier relative to the Earth. By controlling the precession of the gyroscopes on the stabilized platform according to the magnitude of the three-axis directional component of the angular velocity of rotation in the geographic coordinate system, stable tracking of the stabilized platform can be achieved.

The component expressions of rotation angle velocity in the local geographical coordinate system caused by carrier motion are as follows:

$$\omega_E = -\frac{V_N}{R} = \frac{V_{IN}}{R}$$

$$\omega_N = \frac{V_E}{R} + \omega_e \cos \varphi = \frac{V_{IE}}{R}$$

$$\omega_T = \frac{V_E}{R} \tan \varphi + \omega_e \sin \varphi = \frac{V_{IE}}{R} \tan \varphi \qquad (8.5)$$

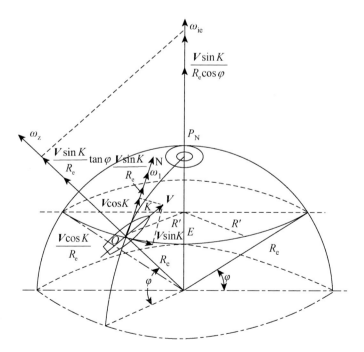

Fig. 8.3 Rotational angular velocity of the local geographic coordinate system caused by vehicle motion

where:

V_E, V_N: the eastward and northward component of the linear velocity of the carrier relative to the geographical coordinate system.

V_{IE}, V_{IN}: the eastward and northward component of the linear velocity of the carrier relative to the inertial space (geocentric inertial coordinate system).

The attitude of the carrier can be measured by the angle sensors installed on the three axes of the stable platform when it has tracked the local geographical coordinate system stably.

Calculation of Velocity

Eastern and northward accelerometers installed on the stabilization platform are used to measure the acceleration of the vehicle in the east–west direction and north–south direction. However, in fact, according to the equivalent effect of gravitation and acceleration in general relativity theory, the accelerometer measurement will include inertial force and gravitation, which is difficult to distinguish from each other. The combination of them measured by the accelerometer is called the specific force.

$$\text{Specific force} = \text{inertial force} - \text{gravitation} \qquad (8.6)$$

The gimballed inertial navigation system adopts stable platform to keep the accelerometer in a horizontal state, thereby achieving isolation of attraction containing gravity. Once there is an error in the platform level, it will lead to errors in acceleration measurement. When calculating carrier velocity based on measured acceleration, it is necessary to fully consider the navigation coordinate system where the velocity located. Once the navigation coordinate system is a non-inertial coordinate system, the motion influence of the navigation coordinate system must be compensated. The harmful acceleration caused by the earth rotation and ship motion in the acceleration output information needs to be compensated by computer, so that the accurate horizontal acceleration can be obtained. The east and north speed components of the ship can be obtained by computer integration.

According to Newton's mechanics principle, when the reference coordinate system (e.g., the earth) rotates at a constant angular rate:

$$\text{Inertial acceleration} = \text{relative acceleration} + \text{implicated acceleration}$$
$$+ \text{Coriolis acceleration} \qquad (8.7)$$

Therefore, by introducing Formula (8.7) into Formula (8.6), we can obtain (8.8):

$$\text{Specific force} = \text{relative acceleration force} + \text{Coriolis acceleration force}$$
$$+ \text{implicated acceleration force} - \text{gravitation} \qquad (8.8)$$

On Earth, gravity is the combination of centrifugal acceleration and gravitation. In summary, if the specific force measured by the accelerometer is $\overline{f}$, the relative acceleration of the vehicle relative to the reference frame is $\dot{\overline{V}}_r$, the angular velocity of the earth's rotation is $\overline{\omega}_{ie}$, the angular velocity of the rotation of the vehicle relative to the earth is $\overline{\omega}_{eb}$, and the local gravity is $\overline{g}$.

$$\frac{\overline{f}}{m} = \dot{\overline{V}}_r + (2\overline{\omega}_{ie} + \overline{\omega}_{eb}) \times \overline{V}_r - \overline{g} \tag{8.9}$$

The above formula is expanded into three-dimensional component form, and the east and north accelerations of vehicles a_E and a_N are assumed to be:

$$f_E = a_E - \left(2\omega_{ie}\sin\varphi + \frac{V_E}{R}\tan\varphi\right)V_N + \left(2\omega_{ie}\cos\varphi + \frac{V_E}{R}\right)V_\varsigma$$

$$f_N = a_N + \left(2\omega_{ie}\sin\varphi + \frac{V_E}{R}\tan\varphi\right)V_E + \frac{V_N}{R}V_\varsigma$$

$$f_\varsigma = a_\varsigma - \left(2\omega_{ie}\cos\varphi + \frac{V_E}{R}\right)V_E - \frac{V_N}{R}V_N + g \tag{8.10}$$

For ships and vehicles moving on the Earth surface, $V_\xi \approx 0$. Therefore, it can be obtained from the above:

$$a_E = f_E + \left(2\omega_{ie}\sin\varphi + \frac{V_E}{R}\tan\varphi\right)V_N$$

$$a_N = f_N - \left(2\omega_{ie}\sin\varphi + \frac{V_E}{R_N}\tan\varphi\right)V_E \tag{8.11}$$

According to Formula (8.11), a_E and a_N are obtained. Then, the eastward and northward velocities V_E and V_N at time k can be calculated according to (8.4).

Calculation of Longitude and Latitude

The velocities V_E and V_N are integrated to obtain the displacement of the carrier. When using longitude and latitude as the positioning parameters, the displacement is added with the initial position λ_0, φ_0 to obtain the longitude and latitude λ, φ of the carrier. When the Earth is regraded as sphere whose radius is R, the longitude and latitude λ and φ of the carrier can be obtained from the following Formula (8.12):

$$\varphi(t) = \frac{1}{R}\int_0^t V_N dt + \varphi_0$$

$$\lambda(t) = \frac{1}{R}\int_0^t V_E \frac{1}{\cos\varphi} dt + \lambda_0 \tag{8.12}$$

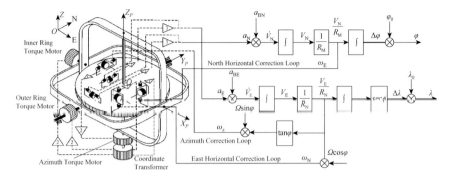

Fig. 8.4 Block diagram of horizontal north pointing semi-analytic INS

where λ_0 and φ_0 are the initial longitude and latitude of the vehicle.

The formulas above are the basic principles of INSs. The block diagram of the horizontal north pointing semi-analytic INS is drawn in Fig. 8.4. The INS is based on Newton mechanics [25]. Linear motion and angular motion relative to inertial space measured by accelerometer and gyroscope. Therefore, accelerometers and gyroscopes are collectively called "inertial sensors".

8.2.2 Gimballed Inertial Navigation

Basic Structure of the Gimballed Inertial Navigation System

According to whether the inertial stabilization platform is adopted or not, inertial navigation systems can be divided into two categories: Gimballed Inertial Navigation System (GINS) and Strapdown Inertial Navigation System (SINS). SINS is such a system in which inertial sensors are directly installed on the vehicle, in which the concept of the navigation platform is replaced by the "mathematical platform" established by computers.

Different types of INSs have different components. Even for the same type of INSs installed on different carriers, their composition will be different. In most cases, the INS consists of the following functional components:

(1) Main instrument: This is also known as the inertial platform, which is the core of the INS. It consists of an inertial platform, shock absorber, temperature control system and electrical components.
(2) Navigation computer: This is mainly used for calculating navigation parameters and command signals applied to gyro torquers. Digital computers are important component of INSs, which accomplishes all calculations and provides control information and data output.

(3) Control and display device: The console is used to operate and control the INS. It includes a working state switch, initial data setting, output data display and fault alarm indication.

(4) Power supply device: power supply is used to supply AC/DC power for electrical components, inertial components and various loops in INSs with high performance requirements.

(5) Electronic equipment cabinet: This includes stable circuit amplifier, accelerometer circuit amplifier, start-up circuit amplifier, inertial components and temperature control circuit of the platform.

(6) Signal transmitter and peripheral equipment: This part is different according to the different requirement of INS applied to various vehicles. Marine INS are often equipped with heading, pitch and rolling transmitter systems.

Classification of Gimballed Inertial Navigation

According to the platform coordinate system established by the stabilized platform, GINS can be divided into three kinds:

(1) **Analytic GINS**

The analytic GINS has three-axis gyro stabilized platform, which is stable in inertial space (system i), so it is also called space stabilized INS.

Three mutually perpendicular gyroscopes and accelerometers are installed on the stabilized platform. The accelerometer measures acceleration of the vehicle in the inertial coordinate system and the gravitation component. When calculating the velocity and position of the vehicle in the inertial coordinate system, the acceleration of the vehicle can be obtained by calculating and eliminating the gravitation influence without modifying the influence of the Earth rotation and the motion of the vehicle. For vehicles moving near the Earth surface, when calculating the velocity and position of the vehicle in the Earth coordinate system or geographic coordinate system, the velocity and position of the vehicle in the inertial coordinate system must be transformed to obtain the velocity, longitude and latitude relative to the Earth coordinate system or geographic coordinate system. Compared with the local horizontal INS, the spatial orientation taken by the platform cannot separate the acceleration of motion from gravity. The data measured by the accelerometer must be analyzed and calculated by computer to obtain the velocity and position of the vehicle, so it is also called the analytic INS.

The platform structure of analytic GINS can be simplified, but due to the need to solve the problems of gravity correction and coordinate conversion, the calculation is heavy. Analytic GINS is applicable to intercontinental missiles, space probes and other carriers flying far away from the Earth.

(2) Semi-analytical GINS

The two horizontal axes of the three-axis stabilized platform of the semi-analytical INS are always in the local horizontal plane, and the vertical axes coincide with the ground vertical. The azimuth refers to the north or a certain direction. This type is also known as the local horizontal INS. Semi analytical INS have the following types.

a. North-pointing INS

The inertial stabilization platform of the north-pointing INS tracks and stabilizes in the local geographic coordinate system, that is, the platform horizontally points to north. It is also called semi-analytical north-pointing INS. North-pointing INS is the most common GINS and suitable for aircraft, ship, combat vehicle and other vehicle moving near the earth surface.

b. Free azimuth INS

The platform of the free azimuth INS is stable in the horizontal plane, while the azimuth axis is not controlled and stable in the inertial space.

c. Wandering azimuth INS

The platform of the wandering azimuth INS is stable in the horizontal plane, and the azimuth axis is controlled by the earth rotation angle speed ω_{ie}, so that the platform rotates around the azimuth axis in space with ω_{ie}. If on the stationary base, the north-pointing INS and the wandering azimuth INS are the same.

In the semi-analytical GINS, gyroscopes and accelerometers are mounted on stable platform. Therefore, the acceleration measured by the accelerometer is the horizontal and vertical component of the carrier relative to inertial space. Because the platform keeps horizontal, the accelerometer output signal does not contain the component of gravity g but contains the harmful acceleration caused by the rotation of Earth and the movement of vehicle. Therefore, the velocity and position of the carrier relative to the earth can only be calculated by integration after the harmful acceleration caused by Earth's rotation and carrier speed is eliminated. Since the vertical acceleration of ships and other vehicles is usually small, the vertical channel accelerometer can be omitted to simplify the calculation of harmful acceleration and system calculation. The most commonly used navigation coordinate system is the local horizontal coordinate system, especially the local geographic coordinate system, because the longitude and latitude calculation in this coordinate system is the most direct and simplest.

(3) Geometric GINS

There are two platforms in this type of INS: one is used to install gyroscopes, which is stable in inertial space; the other is used to install accelerometers, which is stable in the geographic coordinate system (that is, two horizontal axes, one pointing east, one pointing north, and they are always in the horizontal plane). The rotation axis between the two platforms is rotated at the angular velocity of the Earth rotation. At the starting point, the direction of the rotation axis should be adjusted to be parallel to that of the Earth self-rotation axis. Because the platform with the accelerometers tracks the gravity direction and the platform with the gyroscopes is stable in inertial space, this system is also called the gravity INS. According to the geometric relationship between the two platforms of the system, the longitude and latitude of the vehicle can be determined. Therefore, it is called a geometric INS. It has high precision, can work for long time, and has a small amount of calculation, but the structure of the platform is complex. It is mainly used for navigation and positioning of submarines.

Basic Characteristics of Inertial Navigation

Because the INS calculates the acceleration of the vehicle relative to the geographic coordinate system by measuring and calculating the acceleration, the linear velocity is obtained by integrating the acceleration once, and the position information is obtained by integrating the acceleration twice [26]. Therefore, INS is actually a dead reckoning system. Its main characteristics are as follows:

(1)   Capable of working independently and continuously for long time. INS can work independently without receiving any external information. It has good concealment and is not affected by geographical, meteorological and other external environment and time constraints. It can be widely used in many fields, such as aviation, aerospace, marine and land navigation. Especially in underwater navigation, INS is the most important navigation equipment.
(2)   Providing complete vehicle navigation information. It can provide vehicle navigation parameters such as heading, pitch, roll, east speed, north speed, longitude and latitude with good continuity and low noise.
(3)   The positioning error accumulates and diverges with time. The main reason is the divergence of the longitude error and the periodicity of the latitude error. The overall effect is that the positioning error of INS will increase with time. Therefore, the positioning accuracy of INS must specify the time range, for example 1 nautical mile/24 h. For applications where the working time exceeds the designed time duration, the errors need to be corrected periodically, which is called inertial navigation readjustment, to maintain long-term positioning performance.
(4)   The errors of output parameters, such as heading, attitude and velocity, are periodic. It mainly includes 84.4 min of Schuler periodic oscillation, Foucault periodic oscillation related to geographic latitude and 24 h of Earth periodic oscillation. When applied to systems with different working time durations, the error characteristics of INS are different. For example, when used in short-time applications, such as air-to-air missiles, because the working time is much lower than the Schuler period, the error reflects the characteristics of approximately linear growth. When applied to submarines that work for several days, all the oscillation errors of the INS will appear. Because of the application of random errors of inertial components, system errors (such as course errors) will also diverge, and the velocity damping technology is needed to suppress them.

(5) INS has long start-up time. To reduce the initial alignment error of the platform, the system adopts self-alignment methods such as coarse alignment and precise alignment, so the start-up time is long. In the high-precision application field, the start-up process also needs to wait for the gyroscope error characteristics to be stable and can be accurately calibrated, the start-up time is always more than 1 day. For some time-demanding applications, the transfer alignment method is often used to achieve rapid start-up by using higher precision inertial navigation (main INS) speed and attitude data.

(6) The precision of the INS mainly depends on the precision of the gyroscope and accelerometer, the errors of the initial position and velocity, and the initial alignment error of the platform. The performance of the gyroscope determines the positioning accuracy of the INS.

(7) There are differences in the characteristics of inertial technology in different application fields. The characteristics of the aerospace field are high reliability, high accuracy, long endurance, and small size. The marine field emphasizes high reliability, high precision, and long endurance. The aviation industry emphasizes fast start-up, small size, high maneuverability, and strong environmental adaptability. The field of land and petroleum geological exploration and measurement emphasizes fast start-up, small size, and good environmental adaptability. The field of robotics requires fast start-up, small size, light weight, and so on.

8.3 Principle of Strapdown Inertial Navigation

The mature application of solid-state gyroscopes such as ring laser gyroscopes (RLGs), fiber optic gyroscopes (FOGs) and microelectromechanical system (MEMS) gyroscopes has accelerated the development of strapdown inertial navigation technology. The strapdown inertial navigation system (SINS) has developed rapidly in the fields of aeronautics and aerospace since the mid-1970s. They have been widely used and are gradually replacing the GINS.

8.3.1 Brief Introduction of SINS

SINS and GINS are basically the same in configuration. They are mainly composed of Inertial Measuring Unit (IMU), navigation computer and display device. Figure 8.5 shows the block diagram of the SINS. The combination of gyroscopes and accelerometers is commonly called inertial combination. The pointing of the three input axes of gyroscopes and accelerometers should be strictly orthogonal, and the IMU should be installed directly on the vehicle in full accordance with the body coordinate system of the vehicle.

SINSs have no stable platform but directly connect inertial combination to the vehicle, so attitude and heading cannot be directly measured by IMU. To obtain

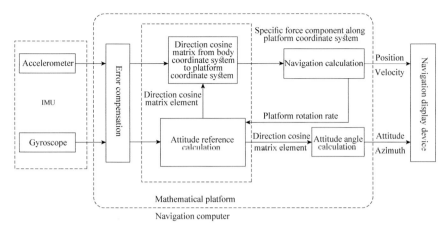

Fig. 8.5 Block diagram of the SINS

the heading and attitude, the output of the gyroscope and accelerometer must be processed in the navigation computer. The attitude matrix of the body coordinate system relative to the geographic coordinate system can be obtained by calculation, and then a 'mathematical platform' in the computer can be established with the heading and attitude calculated. With the help of the 'mathematical platform', the speed and position of the vehicle can be calculated.

SINS can provide position, speed and attitude information, while the strapdown Attitude and Heading Reference System (AHRS) can only provide heading and attitude information. Although there are differences in functions between SINS and AHRS, their basic structures are the same, and their principles are similar and closely connected.

Compared with the stabilized gyrocompass (Chap. 7.3), the AHRS requires high real-time and continuity. Generally, shock absorber is always not installed to improve the response frequency. It has short response time and high frequency of data output. Therefore, AHRS requires high bandwidth, minor data delay and high accuracy. In different applications, AHRSs can be configured differently, but the different requirements of the jitter effect, initial alignment, quasi-instantaneous start-up and structural resonance should be considered (Fig. 8.6).

At present, many companies around the world produce RLG gyrocompass and FOG gyrocompass, such as the United States Kearfott, Honeywell, Northrop Grumman, L-3 Communications company, French Thales, Sagem, IXSea company, German iMAR company, and Russian Astrophysika company (Fig. 8.7).

Fig. 8.6 Nautical -IIG
electric compass, with
permission from [CSSC
Marine Technology]

Fig. 8.7 Conch-90 INS,
with permission from [CSSC
Marine Technology]

8.3.2 Principle of Strapdown Inertial Navigation

The Solution Process of SINS

Unlike GINS, there is no stable platform in SINS, so the specific force measured by
the accelerometer must be projected to the navigation coordinate system through the
attitude matrix firstly to complete the subsequent acceleration calculation, velocity
calculation and position calculation. For ease of understanding, discuss the following
example [27].

For the convenience of the following discussion, it is assumed that the car with a navigation system is limited to a single plane, and its schematic block diagram is shown in Fig. 8.8.

Two accelerometers and a uniaxial rate gyroscope are rigidly mounted on the car base. The sensitive axes of the accelerometer (Fig. 8.9) are perpendicular to each other and are consistent with the axes of the car in the motion plane, represented as x_b and z_b. The sensitive axis of the gyroscope (y_b) is installed perpendicular to the accelerometer two sensitive axes, which measure rotation around the axis perpendicular to the motion plane. Assuming navigation in the reference coordinate system represented by x_i and z_i, the relationship between the vehicle body coordinate system and reference coordinate system is shown in Fig. 8.9. In the figure, θ represents the angular displacement between the reference coordinate system and the body coordinate system. Attitude θ can be determined from the integral of the angular rate ω_{yb} measurable by the gyroscope, as shown in Fig. 8.8:

$$\dot{\theta} = \omega_{yb} \tag{8.13}$$

After obtaining θ, the specific force of accelerometer-measured f_{xb} and f_{zb} can be projected to the reference coordinate system OX_iZ_i.

$$\begin{bmatrix} f_{xi} \\ f_{zi} \end{bmatrix} = \begin{bmatrix} \cos\theta & \sin\theta \\ -\sin\theta & \cos\theta \end{bmatrix} \begin{bmatrix} f_{xb} \\ f_{zb} \end{bmatrix} \tag{8.14}$$

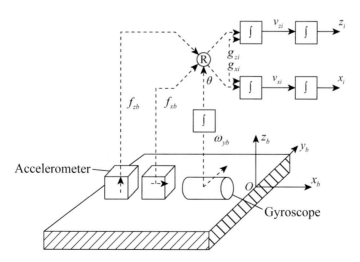

Fig. 8.8 Principle block diagram of two-dimensional SINS

Fig. 8.9 Reference
coordinate system of
two-dimensional

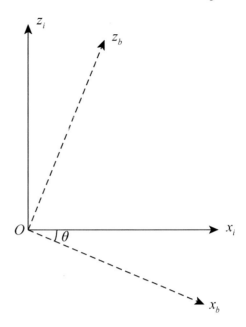

The attitude conversion matrix in the above formula is the key to the strapdown system, which is used to project the specific force of the accelerometer from its sensitive axes to the reference coordinate system axes. This is different from the gimbaled system. Similar to GINS, the relative acceleration of the object relative to the reference frame can be calculated by Formula (8.9) through specific forces f_{xi} and f_{zi}. The relative velocity and position can be obtained by integrating the relative acceleration twice.

In summary, the SINS solution-solving process is summarized as follows:

(1)   Measure the angular velocity of the vehicle relative to the inertial space by the gyroscope and calculate the attitude matrix;
(2)   Project the specific force measured by the accelerometer to the reference coordinate system by the attitude matrix;
(3)   Analyze vehicle motion and calculate relative acceleration;
(4)   Integrate the relative acceleration to obtain the velocity update;
(5)   Integrate the relative velocity to obtain the position update.

Attitude Update

The digital platform of SINS uses the angular velocities measured by strapdown gyroscopes to calculate the attitude matrix, extracts the attitude and heading information of the vehicle from the elements of the attitude matrix, uses the attitude matrix to transform the output of the accelerometer from the body coordinate system to the navigation coordinate system, and then carries out navigation calculations. There are four main methods for describing the attitude relationship of the moving coordinate system relative to the reference coordinate system: the Euler angle method (also known as the three-parameter method), the quaternion method (also known as the four-parameter method), the direction cosine method (also known as the nine-parameter method), and the equivalent rotation vector method. Since the directional cosine matrix method is easiest to understand, the attitude updating algorithm of the directional cosine matrix is introduced here.

1. Directional cosine matrix differential equation

The relation between the relative derivative and absolute derivative of the vector is

$$\left.\frac{d\boldsymbol{r}}{dt}\right|_n = \left.\frac{d\boldsymbol{r}}{dt}\right|_b + \boldsymbol{\omega}_{nb} \times \boldsymbol{r} \tag{8.15}$$

When the amplitude of vector $\boldsymbol{r}$ is unchanged, $\left.\frac{d\boldsymbol{r}}{dt}\right|_b = 0$, put this formula in the geographic coordinate system (n), and represented in the form of a matrix as follows:

$$\dot{\boldsymbol{r}}^n = \left[\boldsymbol{\omega}_{nb}^n \times\right]\boldsymbol{r}^n \tag{8.16}$$

In the formula:

$$\left[\boldsymbol{\omega}_{nb}^n \times\right] = \boldsymbol{\omega}_{nb}^{nK} = \begin{bmatrix} 0 & -\omega_{nbx}^n & \omega_{nby}^n \\ \omega_{nbz}^n & 0 & -\omega_{nbx}^n \\ -\omega_{nby}^n & \omega_{nbx}^n & 0 \end{bmatrix} \tag{8.17}$$

Formula (8.17) is an antisymmetric matrix of the component of the angular velocity of rotation of the body coordinate system relative to the geographic coordinate system along the axis of the geographic coordinate system, usually expressed by symbols $\boldsymbol{\omega}_{nb}^n \times$ or $\boldsymbol{\omega}_{nb}^{nK}$. Each element in $\boldsymbol{\omega}_{nb}^n \times$ is the rotation angular velocity of the body coordinate system relative to the navigation coordinate system and can be obtained by compensating the rotation angular velocity of the reference system from the gyro measurement data.

The vector transformation formula is as follows:

$$\boldsymbol{r}^n = C_b^n \boldsymbol{r}^b \tag{8.18}$$

Derivative on both sides of the formula, and because $\dot{\boldsymbol{r}}^b = 0$, then get:

$$\dot{\boldsymbol{r}}^n = \dot{C}_b^n \boldsymbol{r}^b + C_b^n \dot{\boldsymbol{r}}^b = \dot{C}_b^n C_n^b \boldsymbol{r}^n \tag{8.19}$$

Compared with Formula (8.16):

$$\dot{C}_b^n = \left[\boldsymbol{\omega}_{nb}^n \times\right] C_b^n \tag{8.20}$$

The fourth-order Runge–Kutta method can be used to solve the above equation, and the attitude update matrix C_b^n of the vehicle can be obtained.

2. **Comparison of attitude update algorithms**

The comparison of the above four methods is shown in Table. 8.1.

Table. 8.1 Comparison of attitude update algorithms

Algorithms	Advantage	Disadvantage	Notes
Euler angle	The heading, pitch and roll are directly calculated through the differential equation of Euler angles. The physical concept of Euler angles differential equation is clear, intuitive and easy to understand. There is no need for Orthogonalization in the solution process	Trigonometry is included in the equation, which brings difficulties to real-time computing. When the pitch angle approaches 90°, the equation degenerates	Suitable for situations where there is little change in horizontal attitude, not suitable for attitude determination of full attitude motion carriers
Direction cosine	Solving the attitude matrix differential Equation avoids the problem of equation degeneration in the Euler angles method, and can work in all attitude	The attitude differential equation contains nine position Linear differential equation, which is computationally expensive and difficult to real-time computing calculate	Complex and impractical solution
Quaternion	The attitude is updated by calculating the attitude Quaternion. Need to solve differential equations with 4 unknown quantities, with less computational complexity than directional cosine, simple algorithm, and easy operation	Essentially, the single sample algorithm in the rotation vector method does not compensate enough for non-commutative errors caused by finite rotation	Only suitable for attitude calculation of low dynamic carriers, algorithm drift in attitude calculation can be very severe for high dynamic carriers
Rotation vector	By calculating the quaternion of the attitude, attitude updates can be achieved, and multiple sub sample algorithms can be used to effectively compensate for non commutative errors. The algorithm relationship is simple and easy to operate		Especially suitable for posture updates of carriers with frequent and intense angular excitations or severe angular vibrations

8.3.3 Types of SINSs

The hardware composition of SINS is basically the same, but according to the different navigation coordinate systems, the control scheme is also different. This is because according to the working principle of the accelerometer, for SINSs, the specific force measured by the accelerometer should be $\overline{f}_{ib}^{b}$, which is the specific force of the vehicle relative to the inertia space expressed in the body coordinate system. It needs to be converted to obtain the relative acceleration to the navigation coordinate system. According to the working principle of the gyroscope, the angular velocity measured by the gyroscope should be $\overline{\omega}_{ib}^{b}$, which is the angular velocity of the body relative to the inertia space expressed in the body coordinate system and cannot be directly used for $\omega_{nb}^{n} \times$ in Formula (8.21). It needs to be further calculated to obtain the angular velocity of the vehicle relative to the reference coordinate system. Based on Newton's kinematics analysis, the following is shown:

Gyroscope measurement = angular velocity of the vehicle relative to the reference system + angular velocity of the reference system relative to the inertial space.

Similar to GINS, SINS also has analytical style, semianalytical style and geometric style. The common coordinate systems used in SINSs are the inertial coordinate system, earth coordinate system and geographic coordinate system.

SINS Taking Inertial Coordinate System as Navigation Coordinate System

The reference coordinate system of SINS is inertial system (i system) as the navigation coordinate system, then there are:

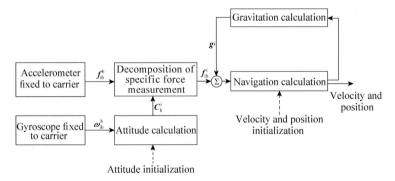

Fig. 8.10 Flow chart of SINS taking the inertial coordinate system as the navigation coordinate system

$$\begin{cases} \overline{\boldsymbol{\omega}}_{ib} = \overline{\boldsymbol{\omega}}_{ib} \\ \overline{\boldsymbol{a}}_{ib} = \overline{\boldsymbol{a}}_i = \overline{\boldsymbol{f}}_{ib} + \overline{\boldsymbol{G}} \end{cases} \tag{8.21}$$

That is, the angular velocity measured by the gyroscope (the right side of equation $\overline{\boldsymbol{\omega}}_{ib}$) is the attitude change rate of the vehicle (the left side of equation $\overline{\boldsymbol{\omega}}_{ib}$). The relative acceleration of the vehicle is the specific force measured by the accelerometer compensated by the gravitation vector (the relative acceleration is the absolute acceleration under the inertial coordinate system). The velocity of the vehicle can be obtained by integrating the acceleration of the vehicle once, and the position of the vehicle can be obtained by integrating the velocity of the vehicle again. Its solution process is shown.

It is easy to solve when INS takes the inertial system as the navigation coordinate system, but its attitude, velocity and acceleration are relative to inertial space, so it is generally used in spacecraft flying away from the earth. Once applied to the near-surface vehicle, the corresponding attitude projection transformation is needed.

SINS Taking the Earth Coordinate System as the Navigation Coordinate System

The angular velocity measured by the gyroscope includes the rotation angular velocity of the earth and the rotation angular velocity of the vehicle relative to the earth. Only the latter can be used to solve the attitude matrix when the earth coordinate system is chosen as the navigation coordinate system. For accelerometers, because the reference coordinate system (earth) rotates at a constant angular velocity, the specific force is compensated by gravitation and extracted with Coriolis acceleration and translational acceleration to obtain the relative acceleration, which can be used in the velocity update and position update equations.

$$\begin{cases} \overline{\boldsymbol{\omega}}_{eb} = \overline{\boldsymbol{\omega}}_{ib} - \overline{\boldsymbol{\omega}}_{ie} \\ \overline{\boldsymbol{a}}_{eb} = \overline{\boldsymbol{f}}_{ib} - 2\overline{\boldsymbol{\omega}}_{ie} \times \overline{\boldsymbol{v}}_{eb} + \overline{\boldsymbol{g}} \end{cases} \tag{8.22}$$

The first formula of (8.22) can be deduced directly from relative motion analysis. The second formula of (8.22) can be deduced according to Formula (8.9). Let the angular velocity of the navigation coordinate system relative to the Earth coordinate system be 0, then obtain the answer. The velocity of the vehicle can be obtained by integrating the relative acceleration of the vehicle once, and the position of the vehicle can be obtained by integrating the velocity of the vehicle again. Its solution flow is shown in Fig. 8.11.

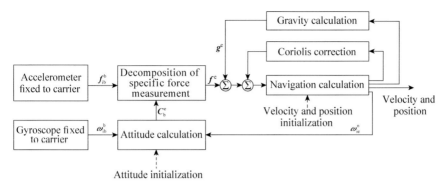

Fig. 8.11 Flow chart of SINS taking earth coordinate system as navigation coordinate system

When using the Earth coordinate system as the navigation coordinate system, the calculation is relatively complicated because of the rotational angular velocity in the reference coordinate system. However, the attitude, velocity and acceleration calculated based on the Earth coordinate system are all relative to the Earth, which is suitable for vehicles such as Earth satellites and other Earth-orbiting motions.

SINS Taking Geographic Coordinate System as Navigation Coordinate System

When choosing the local geographic system as the navigation coordinate system, the rotation of the earth and the rotation of the geographic coordinate system relative to the earth must be taken into account at the same time.

$$
\begin{cases}
\overline{\omega}_{nb} = \overline{\omega}_{ib} - (\overline{\omega}_{ie} + \overline{\omega}_{en}) \\
\overline{a}_{nb} = \overline{f}_{ib} - (2\overline{\omega}_{ie} + \overline{\omega}_{en}) \times \overline{v}_{eb} + \overline{g}
\end{cases}
\tag{8.23}
$$

In the formula, $\overline{\omega}_{en}$ is the angular velocity of rotation between the navigation coordinate system and the earth coordinate system. According to the motion analysis of ships, when the navigation coordinate system adopts the ENT coordinate system, the eastward and northward velocities are V_E and V_N, respectively:

$$
\begin{cases}
\omega_{enx}^{n} = -\dfrac{V_N}{R} \\
\omega_{eny}^{n} = -\dfrac{V_E}{R} + \omega_e \cos \varphi \\
\omega_{enz}^{n} = -\dfrac{V_E}{R} tg\varphi + \omega_e \sin \varphi
\end{cases}
\tag{8.24}
$$

The velocity of the vehicle can be obtained by integrating the relative acceleration of the vehicle, and the position change of the vehicle can be obtained by integrating the velocity of the vehicle again. The calculation process is shown in Fig. 8.12. The geographic coordinate system is suitable for vehicles moving on the Earth's surface.

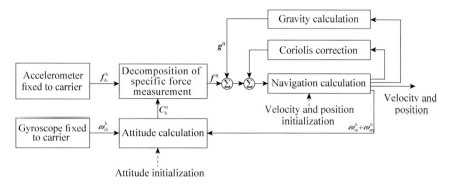

Fig. 8.12 Flow chart of SINS taking geographic coordinate system as navigation coordinate system

8.3.4   Initial Alignment of the SINS

Types of Initial Alignment Methods

The key problem of the initial alignment of SINS is to determine the initial transformation matrix from the body coordinate system to the geographic coordinate system C_b^n (0) with certain accuracy in a given time. The initial alignment of SINS can be roughly classified as follows [28]:

(1) Classified as autonomous alignment and non-autonomous alignment based on whether the external information is used or not.

Autonomous alignment is a method to achieve alignment automatically based on the inertial sensors of the SINS itself. The method of non-autonomous alignment depends on an external reference. Autonomous alignment enhances the autonomy and concealment of SINS, but it requires far more time than non-autonomous alignment. Non-autonomous alignment can rely on external references, which can shorten the alignment time but requires support from external facilities.

(2) Classified as coarse alignment and precision alignment based on different stages.

Coarse alignment can directly estimate the attitude transformation matrix from the body coordinate system to the geographical coordinate system by using the measurement by gravity vector g and the self-rotation rate of the earth ω_{ie} or by means of transfer alignment or optical alignment. The time consumption of coarse alignment is short, and the azimuth and horizontal misalignment angles can be estimated within a certain accuracy range. The dual-vector alignment method introduced in Chap. 5 is a common coarse alignment method. Precision alignment is mainly based on the gyrocompass effect, integrated navigation methods and many other ways to achieve accurate correction and calculation of small misalignment angles between reference coordinate systems and real reference coordinate systems. It provides precise initial conditions for navigation calculation by establishing an accurate initial transformation matrix $C_b^n(0)$. The time of precision alignment is far greater than that of coarse alignment, and the accuracy is higher. The accuracy of precision alignment has a great influence on the accuracy of INS.

(3) Classified as static base alignment and dynamic base alignment based on the base motion state.

The attitude, position and velocity of the vehicle in the ideal state during the alignment of the static base are all 0. At this time, the interference of alignment is small, and the accuracy is high. Dynamic base alignment is more difficult than static base alignment because of the interference angular velocity and acceleration of vehicle motion. To overcome the influence of the motion environment, it is generally necessary to introduce external information and simultaneously restrict the motion of the vehicle.

Precision Alignment Based on the Gyrocompass Effect

INS usually achieves precision alignment by means of "horizontal leveling + azimuth alignment based on the gyrocompass effect", which is called gyrocompass alignment. In the process of horizontal alignment, the horizontal misalignment angle causes velocity error by coupling with gravity, which can be controlled to achieve horizontal alignment. The process of azimuth alignment is similar to that of the gyrocompass and stabilized gyrocompass and achieves azimuth alignment based on the gyrocompass effect. Taking the eastward velocity channel of INS as an example, here briefly introduces the idea of gyrocompass alignment. To deepen the understanding of the gyrocompass alignment from the comparison of GINS and SINS, we first introduce the alignment principle of the east velocity channel of GINS. In fact, this principle is also the basic principle of a stabilized gyrocompass (see Chap. 7). Then, based on this, the alignment principle of SINS is further introduced, which is also the main working principle of the ARHS system.

(1) Alignment of GINS

For convenience of analysis, let the vehicle be stationary and the misalignment angle be $\boldsymbol{\phi} = [\phi_x \ \phi_y \ \phi_z]^T$. The misalignment angle between the geographic coordinate system and the ideal geographic coordinate system tracked by the SINS "mathematical platform" is defined in Fig. 8.13. Let the gyroscope drift be $\boldsymbol{\varepsilon} = [\varepsilon_x \ \varepsilon_y \ \varepsilon_z]^T$, accelerometer bias be $\boldsymbol{\nabla} = [\nabla_x \ \nabla_y \ \nabla_z]^T$, and speed error be $\delta \boldsymbol{V} = [\delta V_E \ \delta V_N \ \delta V_T]^T$.

Under the condition of a stationary base and small misalignment angle, the velocity error and misalignment angle equation of the eastward passage can be approximated as follows:

$$\begin{cases} \delta \dot{V}_E = -\phi_y g + \nabla_x + \dot{V}_E \\ \dot{\phi}_y = -\frac{\delta V_E}{R} + \varepsilon_y \end{cases} \tag{8.25}$$

From (8.25), a block diagram of the eastward horizontal loop is shown in Fig. 8.14.

Fig. 8.13 Definition of
platform misalignment angle

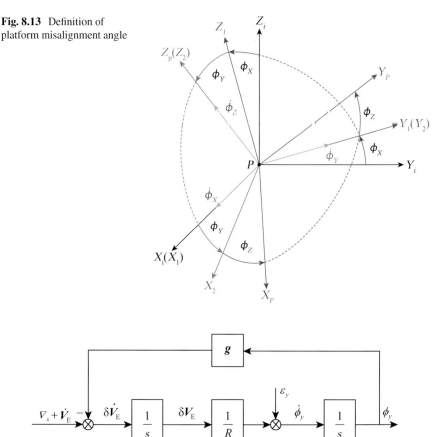

Fig. 8.14 Block diagram of the eastward horizontal loop

According to the principle of automatic control, the eastward channel is critically
stable, and the misalignment angle does not converge. To converge the misalignment
angle, the third-order horizontal leveling loop of the east channel is usually adopted
for gyrocompass alignment (Fig. 8.15).

In the third-order horizontal levelling loop, by introducing node K_1 to achieve
velocity feedback and damping, adding node K_E to adjust the scale coefficient propor-
tionally, and adding the integral node K_U/S to eliminate the constant error caused by
gyro drift and platform azimuth misalignment, the steady-state error of the horizontal
misalignment angle ϕ_y will be $-\nabla_x/g$. The comparison between Figs. 8.14 and 8.15
shows that introducing three nodes, K_1, K_E, and K_U, will change the control of the
angular velocity along the east direction of the platform. Through the same method, it
can control the north channel and azimuth channel, obtain ω_{cE} and ω_{cT}, and achieve
accurate alignment of the inertial platform.

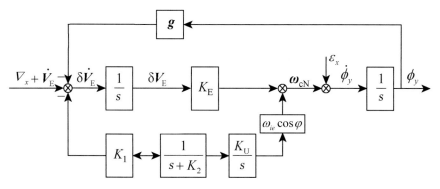

Fig. 8.15 Block diagram of the third-order horizontal leveling loop in the eastern channel

(2) Gyrocompass alignment method of SINS

SINS replaces the real platform with a mathematical platform calculated in the computer. Therefore, the gyrocompass alignment method of GINS can be transplanted into the SINS. That is, the signal flow of stabilized gyrocompass alignment can be used to control the platform movement, which is realized by a mathematical method. The algorithm principle of the mathematical platform is shown in Fig. 8.16.

As seen in Fig. 8.16, $C_b^{\hat{n}}$ is the computed strapdown matrix, and ω_{ib}^b and f^b are the measurements of the gyroscopes and accelerometers, respectively. Ω_{ib}^b is the anti-symmetric array of ω_{ib}^b, and $\Omega_{ie}^{\hat{n}}$ is the anti-symmetric array of the projection of the earth rotation angular rate $\omega_{ie}^{\hat{n}}$ in the computational navigation coordinate system, where $\Omega_c^{\hat{n}}$ is the anti-symmetric array of the mathematical platform modified angular rate vector ω_c projected in the computational navigation coordinate system. $f^{\hat{n}}$ is the output of the transformation of f^b by calculating the strapdown matrix $C_b^{\hat{n}}$.

The alignment rule of the GINS can be transplanted to the SINS if ω_{cE}, ω_{cN} and ω_{cT} are equivalent to the control angular velocity ω_c that is mathematically introduced. As shown in Fig. 8.17, by processing the accelerometer specific force, constructing the modified angular rate vector ω_{cN}, and introducing it into the attitude update

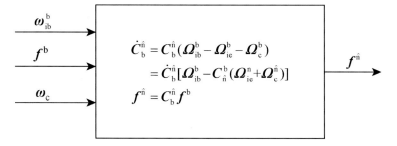

Fig. 8.16 Algorithm principle of the SINS mathematical platform

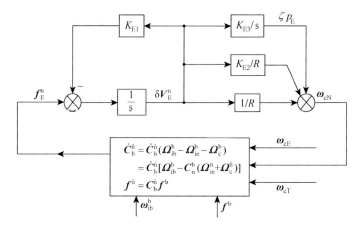

Fig. 8.17 Principle diagram of the horizontal alignment of the eastern channel

calculation as a control quantity, the misalignment angle can gradually converge and finally reach stability, similar to the stabilized gyrocompass.

Figure 8.18 can also introduce a modified angular rate vector for the northward and azimuth channels $\boldsymbol{\omega}_{cE}$ and $\boldsymbol{\omega}_{cT}$. If it is used as a control variable in the attitude updating calculation, the misalignment angle will converge gradually and finally reach stability, similar to GINS.

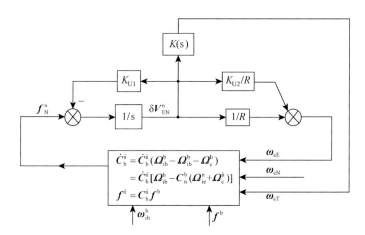

Fig. 8.18 Principle diagram of azimuth alignment

8.3.5 Basic Characteristics of SINS

For GINS and SINS, their working principles are basically the same. The biggest problem with INS is that its positioning error accumulates with time, and the long-term stability of system accuracy is poor. To improve the accuracy of INS, many countries have invested much in the development of high-precision INS. Compared with GINS, SINS often has the advantages of fast start-up, strong mobility adaptability, small size, good fitness, all-solid-state design and high reliability. Generally, the characteristics of SINS are as follows:

(1) Because IMUs are directly attached to the vehicle, they measure the inertial linear acceleration along the axes of the body coordinate system and the angular rate of rotation around the axes of the body system, which is different from GINS. In GINS, the output of the gyroscope and accelerometer is used to stabilize the platform coordinate system in the navigation coordinate system, and the accelerometer is installed on the platform. Therefore, the accelerometer can directly measure the acceleration of the vehicle along the axis of the platform system (navigation coordinate system). In addition, because the stabilized mechanical platform is stable in the navigation coordinate system, when the navigation coordinate system is a geographic coordinate system, each axis of the platform can directly output the attitude angle of the vehicle. In SINS, the output of the accelerometer must be converted to a navigation coordinate system, and then the navigation parameters can be calculated. The output of the gyroscope is used to establish and modify the mathematical platform (navigation coordinate system) and to calculate the attitude angle. Complex electrical and mechanical platforms are completely replaced by computer software functions, which is the most important feature of SINSs.

(2) Compared with GINS, SINS can provide more navigation and guidance information, such as attitude angular rate and acceleration measurement. The RLG SINS used in Boeing aircraft 757 and 767 can provide up to 35 kinds of information.

(3) Although the form of the error equation is basically the same, the error propagation characteristics of SINS are different from those of GINS with different trajectories.

(4) Because the physical platform is omitted, the size, weight and cost of the SINS are greatly reduced, and the reliability is improved. Because the IMU only plays the role of transferring information and has no feedback control for the motor, it is an open-loop control. All signal processing is realized in the computer, so it is convenient to realize. It is easy to realize miniaturization and low cost.

(5) IMUs are easy to install, maintain and replace and to adopt redundancy configurations to improve the performance and reliability of the system.

(6) The IMU is fixed on the vehicle, which directly bears the vibration and impact of the vehicle, and the working environment is tough. Therefore, inertial components in SINS need to have higher impact and vibration resistance. The dynamic

environment of the vehicle causes great errors in inertial instruments. Inertial instruments, especially gyroscopes, directly measure the angular motion of the vehicle; for example, the maximum angular velocity of a high-performance fighter reaches 400°/s, while the lowest is likely to be as low as 0.01°/h. In this way, the range of the gyroscope is as high as that of the gyroscope with order 108.

(7) The gyroscope of the GINS is installed on the platform and can be tested arbitrarily relative to the acceleration of gravity and the angular velocity of the Earth's rotation, which is convenient for error calibration. The strapdown gyroscope does not meet this requirement, so it is difficult to calibrate. It can only take the IMU off the vehicle and calibrate it in the laboratory. In the course of use, depending on the stability of instrument performance or using external information, calibration can be carried out during vehicle movement.

(8) Because SINS components are directly mounted on the vehicle and directly sensitive to complex motions such as sway and oscillation of the vehicle, the working environment is harsh, and there are many disturbances, so there will be errors such as paddling and coning under highly dynamic conditions. This kind of error will cause large errors when the vehicle is maneuvering, so the real-time compensation of dynamic errors is a problem that must be solved for high-performance SINS.

(9) In SINS, the computational load of the computer is much larger than that of GINS, and the word length and speed requirements of the computer are much higher. Due to the amazing development of computer technology, this requirement has been solved, and 32-bit microcomputers or microprocessors can meet the requirement.

This chapter introduces two kinds of autonomous positioning methods based on the vehicle itself. Compared with direct positioning systems, it does not need to observe external references, is not restricted by weather and geographical conditions, and has strong confidentiality. It can work continuously, has a high update rate and low short-term noise. Because it does not have any optical and electrical connections with the outside world, it is especially suitable for military purposes. It overcomes the disadvantage of observation navigation based on external information. However, there are some limitations in these autonomous navigations: position parameters must be initialized, and errors accumulated resulting in position errors increasing with time. Therefore, after a period of navigation, other means are needed to help the system re-calibrated. In integrated navigation systems, direct positioning measurements are often used to correct the results of dead reckoning navigation and compensate for the error of inertial sensors.

Appendix 8.1: Questions

(1) Through data retrieval, please briefly describe the dead reckoning methods and characteristics used in automobile navigation.

(2) Please briefly describe the basic method of ship dead reckoning and the main influencing factors that need to be considered.

(3) Please briefly describe the basic principles, features and application fields of inertial navigation. Considering the characteristics of INS, please analyze the importance of inertial navigation.

(4) What are the different technical characteristics and difficulties of long operation-time INSs and short operation-time INSs?

(5) What are the types of gimballed inertial navigation systems, and what are their application characteristics?

(6) What are the core components of the INS? Summarize the connections and differences of inertial sensor, inertial combinations, and inertial measurement units? Please briefly describe the types and characteristics of the gyroscopes and accelerometers.

(7) What are the similarities and differences in the system composition, working principle and performance characteristics of gimballed inertial navigation systems and SINSs?

(8) Please briefly describe the working principle of the initial alignment of the INS, compared with the working principle of the gyrocompass and stabilized gyrocompass to see the connection.

Appendix 8.2: The Philosophy Behind the Inertial Navigation

The dead reckoning and inertial navigation principle is the core principle of autonomous navigation. The idea is to calculate accurately the motion parameters by measuring the linear and angular motions of the given carrier in real time. For instance, the carrier position can be determined through recursive calculation or integral operation. More importantly, there are many scientific and philosophical ideas behind such a principle. For example, if the motion parameters of each time interval are calculated accurately, you can calculate the parameters for days and months without any mistakes, which echoes the idea of "a journey of thousand miles begins with a small step (千里之行, 始于足下)". Just as the saying in *Tao-Te Ching* goes, "all great things have small beginnings. (天下大事, 必作于细)" It is the angular speed and acceleration measured each time that determine the overall accuracy of the system.

To put it another way, the system performance is directly determined by the accuracy of inertial components (e.g., the gyro and accelerometer), despite their small size and low power consumption. This cause-and-effect relation can also be observed between one's daily efforts and their ultimate excellence. Even small efforts can make a big difference as long as they are ongoing; meanwhile, it is also important to keep the ongoing efforts on the right track, which makes the right goal most crucial.

In addition, a gyro with high precision and a strong fixed axis is similar to a person with focus. Focus can make people rational and calm and thus keep pursuing meaningful goals, which in turn leads to breakthroughs and innovations.

Still, the mathematical basis of the inertial navigation is the kinematics calculus formula, which reveals the coexistence of "movement" and "stillness" in the world. This dialectical unity touches on the core of ontology and epistemology. Developing such a dialectical way of thinking (e.g., absoluteness vs. relativity) can help us see the world more comprehensively, objectively, and with an open mind.

Chapter 9
Integrated Navigation

Navigation systems have a variety of types and can provide information in different working conditions by various means. Thus, in practice, people seldom rely on a single navigation system but combine different navigation systems to form an integrated navigation system with better comprehensive performance. For example, rockets can use INS/CNS/GPS integrated navigation systems; aircraft can use INS/GPS/TNS integrated navigation systems; and even relatively simple land vehicle navigation systems are also based on GPS, path planning or DR (gyroscope/odometer) technology. Even though its working condition is critical, submarine navigation systems, which mostly rely on the INS, are actually integrated navigation systems that combine INS, log and gravity matching systems. Therefore, from a certain point of view, almost all navigation systems in various applications, such as maritime, land, air and space, are integrated navigation systems. At present, the high-tech war has put forward very high requirements for the autonomy, accuracy, automation and compactness of the navigation system. With the development of technology, increasingly more information resources of navigation systems are available. Integrated navigation is the result of the development of modern navigation theory and technology. Each navigation system has its own unique performance and limitations. Combining several different systems, we can make use of multiple information sources to complement each other and form a multifunctional system with more redundancy and accuracy.

© Science Press 2024   223
H. Bian et al., *Essentials of Navigation*, https://doi.org/10.1007/978-981-99-5636-4_9

9.1   Basic Aspects of Integrated Navigation

9.1.1   Concept of Integrated Navigation

According to the classification of multisensor information fusion theory, integrated navigation multisensor information fusion belongs to fusion at the position-level and attribute-level. The integrated navigation system should receive and process all available navigation information adaptively and fuse the navigation information to provide more accurate information such as position, speed and attitude. According to the requirements of reliability and robustness of the system, the high-precision navigation system must also have strong error tolerance, fault diagnosis and isolation, optimization of the information, and provide redundant navigation information and auxiliary decision making.

9.1.2   Types of Integrated Navigation

INS/GNSS Integrated Navigation System

INS does not rely on external information. It has good concealment and strong anti-interference ability and can provide almost all the navigation parameters needed; it has the advantages of a high data update rate, short-term accuracy and good stability. However, its error accumulates over time, and the initial start-up alignment time is too long, which is fatal for applications that require rapid response and long-time operation accuracy. GNSS is a star-based navigation and positioning system that can provide accurate three-dimensional position, speed and time continuously all the time. However, it has the disadvantages of poor dynamic response ability, poor immunity to signal interference, blocking and poor integrity.

When INS and GNSS are combined, high-precision GNSS information is considered an external measurement to limit the accumulation of INS errors over time, while the short-term stability of INS can solve the problem of GNSS signal loss and cycle jump in the dynamic environment. Therefore, the integrated system not only has the advantages of the two systems but also has much better overall performance with the deepening of the combination level and the enhancement of information fusion. The common integrated methods include loose coupled, tight coupled, ultratightly coupled, closely coupled, cascaded and deep coupled. As a result, the combination of INS/GNSS is considered to be the most ideal integrated navigation.

INS/CNS Integrated Navigation System

Because celestial objects cannot be destroyed and disturbed, INS/CNS integrated navigation systems are autonomous navigation systems. On the one hand, INS can provide CNS with attitude, heading, speed and other data, helping it to obtain position and orientation more accurately and quickly; on the other hand, CNS can also be used to correct INS parameters such as position and heading. Based on the attitude angle error obtained and the Kalman filter, INS inertial sensors and parameter errors can be estimated and compensated to improve its accuracy.

GNSS Integrated Navigation System

At present, GNSS systems include GPS, Glonass, BDS and Galileo. When a single GNSS receiver is used in special occasions such as canyons and forests, the number of visible satellites will be reduced, thus decreasing the positioning accuracy of the system. Furthermore, for military and political reasons, US nonallied users are not guaranteed to use GPS. As a result, people began to study GPS/Glonass/BDS/Galileo integration methods to improve the accuracy and reliability of GNSS. The integrated receiver will receive different GNSS satellite signals simultaneously. Compared to a single GNSS, the number of satellites available for the integrated system has theoretically increased nearly four times, so the number of visible satellites increases under the same conditions, and the reliability of positioning accuracy will be greatly improved.

GNSS/Loran C Integrated Navigation System

The Loran C system is a land-based remote radio navigation system using a solid-state high-power transmitter with a peak transmission power of up to 2 MW. It has large coverage, strong anti-interference ability and high reliability. Loran C and GNSS are independent of each other, and each has its own advantages and disadvantages. As a result, GNSS/Loran C integrated navigation can achieve better performance. There are several ways to combine Loran C and GNSS:

(1) Loran C-based GNSS differential enhancement: the GNSS pseudorange difference-corrected information can be modulated and broadcast with the Loran C signal, and the user can receive and demodulate the Loran C and GNSS signals. Through sample integration, GNSS differential positioning is achieved.
(2) Loran C serves as a pseudossatellite to enhance GNSS: Loran C station is used to send out GNSS satellite navigation information message data, and Loran C station works like a pseudossatellite.
(3) Using GNSS to improve the Loran C positioning accuracy: ASF error is the main problem that influences the accuracy of Loran C, and its value can be calculated and measured by using GNSS information to improve the positioning accuracy of the Loran receiver.
(4) Compensated Loran C and GNSS data fusion: The error of Loran C is much larger than that of GNSS. The accuracy of the ASF-compensated Loran receiver can reach 8–20 m, which is close to that of GNSS. Since they are two independent

navigation systems, the compensated Loran C and GNSS can be combined to obtain better positioning results than the GNSS receiver.

GNSS/DR Integrated Navigation System

GNSS/DR integrated navigation is often used in low-cost applications such as automobile navigation. The DR system cannot provide the initial coordinate and heading of the vehicle, and its position error accumulates and diverges gradually. Thus, it is not suitable for long-term autonomous navigation, so other means are needed to compensate for its accumulated error properly. GNSS/DR integrated navigation systems can fully combine the advantages of the two means. The GNSS absolute position can be the initial value for the DR system and be used to correct the DR parameters. At the same time, the DR system provides continuous navigation results, which can compensate for the random error and breakpoint of the GNSS signal and smooth the positioning trajectory.

9.2 Integrated Navigation Principles

Integrated navigation information fusion is based on the physical model of the system (described by the state equation and observation equation) and the statistical assumption of sensor noise and maps the observation data to the state vector space. The state vector includes a set of state variables of the navigation system, such as position, velocity, angular velocity, attitude and various misalignment deviations. The state variables can be used to accurately describe the state of the system and the motion of the vehicle. The fusion process for a multisensor navigation system is actually the interconnection of sensor measurement data and the estimation of the state vector. The input data of various sensors are first transformed into the same navigation coordinate system through coordinate transformation and unification. The data belonging to the same state are connected. According to the established motion law of the vehicle, the state equation of the system and the mathematical mode of observation quantity, under certain optimal estimation criteria, employing the optimal estimation to best fit the state vector and the observation obtains the best estimation of the state vector. The optimal estimation criteria include the least squares method (LSM), weighted least squares method, least mean square error (LMSE), maximum likelihood (ML), Bayesian criterion, etc. The most commonly used integrated navigation algorithm is the Kalman filter.

9.2.1 Basic Principles of Kalman Filter

Introduction

The basic principle of the Kalman filter can be understood intuitively through a common example.

Taking the target attacked by naval gun as an example, the ship must be able to hit the moving target accurately in distance. Because the target is moving continuously, it is difficult to hit the target if the gun firing is only according to the measured target azimuth and distance. This is because the projectile takes time to hit, and by the time it arrives, the target has already left its original position. Therefore, the crew must predict the target's azimuth and distance based on the observed target movement. Specifically, the crew must continuously track and aim the target, record its data at each moment, and estimate the position of the target for the next moment. However, in the second measurement, it is often found that there is a deviation between the estimated position and the measured position, so the crew can adjust and correct the estimation accuracy at the next point. If the velocity of the target estimation is larger, the estimated position will be ahead of the actual target, and the predicted displacement distance of the target in this direction should be reduced. In contrast, if the velocity of target estimation is smaller, which causes the estimated position to lag behind the actual target, the predicted displacement distance of the target in this direction should increase. After several adjustments, the position estimation and prediction of the target become more accurate until the best effect is achieved.

Let us analyze the above process carefully, and we can find some important steps. After the crew acquires a current measured target position, the crew immediately predicts the target possible position of the next moment, which is called state prediction. Because the prediction is often inaccurate, when the next target observation is obtained, the error of the prediction can be found immediately. The crew will adjust the estimation of the next target position according to this error. This is called the estimation correction. The optimal estimation of the state is achieved gradually by iterating these two steps continuously.

Basic Formula of Kalman Filter

For deterministic systems, given the initial conditions of the system, the exact state of the system in the future can be obtained by solving the differential equation of the system. However, in practice, most of the systems are stochastic dynamic systems, which are affected by various disturbances and noises in the operation process, which bring some uncertainty to their operation status and thus produce various errors. Integrated navigation systems are stochastic dynamic systems. The Kalman filter is the most commonly used estimation algorithm in integrated navigation systems and uses the state space method to establish an accurate state equation and measurement equation of the system. At the same time, it establishes the precise white noise

statistical characteristics of the system and measurement noise [29]. By establishing
a set of real-time recursive algorithms realized by computers, the optimal estimation
of the system state can be achieved according to the observation and measurement
of the system at every moment.

The Kalman filter is a linear minimum variance estimation. It uses the state space
method to filter in the time domain and is suitable for estimating multidimensional
stochastic processes. Kalman filters have many theoretical deduction methods and
different representation methods. Here, the discrete Kalman recursive filtering (DKF)
algorithm is given directly.

First, stochastic discrete linear equations are used to describe the system, and the
state equation and measurement equation of the time system at t_k are described as
follows:

$$X_k = \Phi_{k,k-1} X_{k-1} + \Gamma_{k,k-1} W_{k-1} \tag{9.1}$$

$$Z_k = H_k X_k + V_k \tag{9.2}$$

where X_K is the estimated state vector, W_K is the system noise sequence vector, V_K
is the measure noise sequence vector, $\Phi_{k,k-1}$ is the one-step transfer matrix from
step t_{k-1} to step t_k; $\Gamma_{k,k-1}$ is the system noise driven matrix; H_K is the measurement
matrix; W_K and V_K satisfy the following: $E[W_k] = 0$, $\text{Cov}[W_k, W_j] = Q_k \delta_{kj}$,
$E[V_k] = 0$, $\text{Cov}[V_k, V_j] = R_k \delta_{kj}$, $\text{Cov}[W_k, V_j] = 0$, δ satisfy Dirac function
$\delta_{kj} = \begin{cases} 1, k = j \\ 0, k \neq j \end{cases}$; Q_k is the variance matrix of the system noise sequence, assuming
that it is a nonnegative definite matrix; R_k is the variance matrix of the measurement
noise sequence, assuming that it is a positive definite matrix.

The calculation steps of the DKF are as follows:

One-step prediction of state:

$$\hat{X}_{k,k-1} = \Phi_{k,k-1} \hat{X}_{k-1} \tag{9.3}$$

State estimation:

$$\hat{X}_k = \hat{X}_{k,k-1} + K_k \left(Z_k - H_k \hat{X}_{k,k-1} \right) \tag{9.4}$$

The filter gain matrix:

$$K_k = P_{k,k-1} H_k^T \left(H_k P_{k,k-1} H_k^T + R_k \right)^{-1} \tag{9.5}$$

One-step prediction error variance matrix:

$$P_{k,k-1} = \Phi_{k,k-1} P_{k-1} \Phi_{k,k-1}^T + \Gamma_{k,k-1} Q_{k-1} \Gamma_{k,k-1}^T \tag{9.6}$$

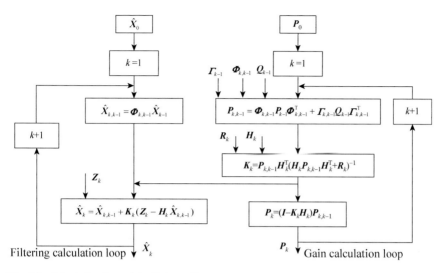

Fig. 9.1 Calculation flow of the Kalman filter

Estimation error variance matrix:

$$\boldsymbol{P}_k = (\boldsymbol{I} - \boldsymbol{K}_k \boldsymbol{H}_k) \boldsymbol{P}_{k,k-1} \qquad (9.7)$$

As long as the initial values $\hat{\boldsymbol{X}}_0$ and P_0 are given, according to the measurement value $\boldsymbol{Z}_k$ of step k the state estimation $\hat{\boldsymbol{X}}_k (k = 1, 2, \ldots)$ of step k can be calculated recursively. The basic algorithm of the discrete Kalman filter shown in Formulas (9.3)–(9.7) can be described by Fig. 9.1.

It is obvious from Fig. 9.1 that the Kalman filter has two calculation loops, namely, the estimation calculation loop (left side) and gain calculation loop (right side). The gain calculation loop is an independent calculation loop, while the estimation calculation loop depends on the gain calculation loop. In a filtering cycle, we can see that the Kalman filter has two processes: time update and measurement update. These two processes use system information and measurement information to update the system state estimation successively.

Formulas (9.3) and (9.6) belong to the process of time renewal, Formula (9.3) explains the method to predict the state estimation of step k based on the state estimation of step $k - 1$, and Formula (9.6) quantitatively describes the quality of this prediction. In the calculation of the two formulas, only the information related to the dynamic characteristics of the system is used, such as the one-step transfer matrix, noise driving matrix and variance matrix of driving noise. From the process of time lapse, these two formulas push the time of state estimation from step $k - 1$ to step k only according to the system's own characteristics. The measurements are not used at any time, so what they describe is the time update process of the Kalman filter. This process is similar to the prediction of target tracking.

The measurement renewal process is mainly described by Formulas (9.4), (9.5) and (9.7). They are mainly used to calculate the correction of the time renewal value, which is determined by the quality of time renewal ($P_{k/k-1}$), quality of measurement information (R_k), the relationship between measurement and state (H_k) and the specific measurement Z_k. It is determined that all these equations revolve around one purpose, that is, to use the measurement Z_k correctly and reasonably. Therefore, this process describes the measurement update process of the Kalman filter, which is similar to the previous target tracking modification.

9.2.2 Basic Principles of Recursive Bayesian Estimation

Bayesian estimation theory is an important branch of mathematical probability theory. Its basic idea is to obtain posterior information by combining prior information of random variables with new observation samples. The prior distribution reflects the cognition of the statistical characteristics of random variable samples before the experiment. With the new observation information of samples, the cognition has changed. The result must be reflected in the posterior probability distribution; it is said that the posterior distribution synthesizes the prior distribution and the information of samples. If the posterior distribution of the previous moment is taken as the basis for solving the prior distribution of the next moment, the recursive Bayesian estimation is formed by iteration and recursion in turn.

In fact, various forms of Kalman filters, unscented Kalman filters and particle filters are all special forms of Bayesian estimation. Bayesian estimation is the basic form of many integrated navigation filtering algorithms and has the inherent unified essence.

Bayesian Formula

Given the occurrence of event B, the conditional probability of event A is expressed by P (A | B). The Bayesian formula can be expressed as follows:

$$P(A_i|B) = \frac{P(B|A_i)P(A_i)}{\sum_{i=1}^{n} P(B|A_i)P(A_i)} \tag{9.8}$$

Recursive Bayesian Estimation

It is assumed that the state equation of the discrete system described by the generalized state space model is as follows:

$$x_{k+1} = f(x_k, w_k) \tag{9.9}$$

$$z_k = g(x_k, v_k) \tag{9.10}$$

Equation (9.9) is the state equation, which describes the state transition probability of the system $P(x_{k+1}|x_k)$. Equation (9.10) is a measurement equation, which describes the transition probability function of state measurement $P(z_k|x_k)$. It is related to the actual measurement noise model. $f: R^{N_x} \Rightarrow R^{N_x}, g: R^{N_x} \Rightarrow R^{N_y}$, w_k and v_k are white noise with unknown statistical characteristics. To simplify the problem, assume that the above system satisfies:

System state follows the first-order Markov process, then $P(x_k|x_{0:k-1}) = P(x_k|x_{k-1})$;

(1) The observed values are independent of the internal state of the system.

Define the sequence of state variables $X_k = [x_0, x_1..x_k]$ and the observation sequence $Z_k = [z_0, z_1...z_k]$. $P(x_k|Z_k)$ is the conditional probability density function of x_k. According to Bayes Formula (9.8), the following results are obtained:

$$
\begin{aligned}
P(x_k|Z_k) &= \frac{P(Z_k|x_k)P(x_k)}{P(Z_k)} \\
&= \frac{P(z_k, Z_{k-1}|x_k)P(x_k)}{P(z_k, Z_{k-1})} \\
&= \frac{P(z_k|Z_{k-1}, x_k)P(Z_{k-1}|x_k)P(x_k)}{P(z_k, Z_{k-1})P(Z_{k-1})} \\
&= \frac{P(z_k|Z_{k-1}, x_k)P(x_k|Z_{k-1})P(Z_{k-1})P(x_k)}{P(y_k|Z_{k-1})P(Z_{k-1})P(x_k)} \\
&= \frac{P(z_k|x_k)P(x_k|Z_{k-1})}{P(z_k|Z_{k-1})}
\end{aligned} \tag{9.11}
$$

There are several conceptual terms in the above formula that must be explained here.

The mathematical meaning of $P(x_k|Z_{k-1})$ is the probability that the system state is x_k at the moment$k$, when the obtained sequence of systematic observations is $Z_{k-1} = [z_0, z_1...z_{k-1}]$ from step 0 to step $k-1$. This is a prior probability density function, which predicts the state of the system at the next moment based on the previous observations.

The mathematical meaning of $P(x_k|Z_k)$ is the probability that the system state is x_k at the present moment k when the obtained sequence of systematic observations is $Z_k = [z_0, z_1...z_k]$ from step 0 to step k. This is a prior probability density function, which predicts the state of the system at the next moment based on the previous observations. Because the system observations z_k are actually the external performance of the system state x_k, it is actually used to analyze the reasons based on the results or to judge the system input probability according to the known system output, so it is called the posterior probability density function. Note that x_k is the true value of the system state that is expected to be estimated. In the sense of maximum a posteriori estimation, the system state estimation is expected to obtain maximum probabilities when x_k appears.

$P(x_k|x_{k-1})$ is the probability that the system state is x_k at the present k moment when the system state is x_{k-1} at the moment $k-1$. It is known as the probability density function of state transition. $P(z_k|x_k)$ is called Likelihood. $P(y_k|Y_{k-1})$ is called Evidence.

The formulas for calculating the denominator of Formula (9.11) are as follows:

$$P(x_k|Z_{k-1}) = \int P(x_k|x_{k-1})P(x_{k-1}|Z_{k-1})dx_{k-1} \tag{9.12}$$

$$P(z_k|Z_{k-1}) = \int P(z_k|x_k)P(x_k|Z_{k-1})dx_k \tag{9.13}$$

Here, the recursive Bayesian estimates are expressed as follows:

Assuming that the posterior probability density $P(x_{k-1}|Z_{k-1})$ at step $k-1$ is given, the prior probability density function at step k can be obtained by time updating:

$$P(x_k|Z_{k-1}) = \int P(x_k|x_{k-1})P(x_{k-1}|Z_{k-1})dx_{k-1} \tag{9.14}$$

After obtaining new observation information z_k at step k, the calculation formula of the posterior probability density function is as follows:

$$P(x_k|Z_k) = \frac{P(x_k|x_k)P(x_k|Z_{k-1})}{\int P(z_k|x_k)P(x_k|Z_{k-1})dx_k} \tag{9.15}$$

The recursive process is shown in Fig. 9.2. The recursive Bayesian estimation process can be further abstracted into two basic steps: prediction and correction.

The problem of state filtering estimation for stochastic systems can be described in the form of Bayesian estimation, which is a very general problem. For linear Gauss stochastic dynamical systems, the optimal solution of state filtering estimation is the Kalman filtering solution.

Kalman filtering is essentially a recursive estimation of the system state under constantly updated observation information. One-step prediction of the state and its covariance can be obtained from the system state equation, and then the correction of one-step prediction under observation can be obtained from the system observation equation. Taking the one-step prediction of the system state and covariance as a

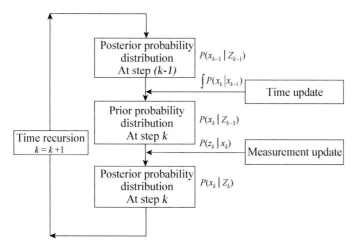

Fig. 9.2 Recursive Bayesian estimation flow

prior distribution, the Kalman filtering process is a recursive Bayesian estimation process to obtain the posterior distribution under the condition of constantly updating observation information. Therefore, the Kalman filtering algorithm can be perfectly unified in the framework of Bayesian estimation theory.

9.2.3 Common Integrated Navigation Architecture

Direct and Indirect Methods

When designing an integrated navigation system, the system equation describing the dynamic characteristics of the navigation system and the measurement equation must be listed first [30]. If the navigation parameters of each subsystem are taken as the states to be estimated, the filtering processing is called direct filtering. If the errors of each subsystem are taken as the state to be estimated, the filtering process is called indirect filtering.

In direct method filtering, the integrated navigation state filter receives the navigation parameters of each navigation subsystem. After filtering calculation, the optimal estimation of the navigation parameter is obtained, as shown in Fig. 9.3.

In indirect filtering, the integrated navigation state filter receives the difference between the output values of multiple navigation subsystems for the same navigation parameters and then estimates the errors by filtering calculation, as shown in Fig. 9.4.

The advantages and disadvantages of direct and indirect filtering are as follows:

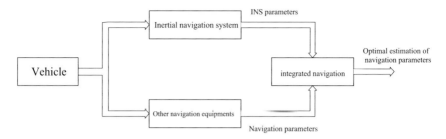

Fig. 9.3   Diagram of direct method filtering for INS-based integegrated navigation system

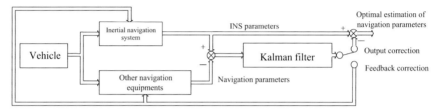

Fig. 9.4   Diagram of indirect method filtering

(1) The model system equation of the direct method directly describes the dynamic
     process of the navigation parameters of the system, which can more accurately
     reflect the evolution of the real state; the model system equation of the indirect
     method is the error equation, which is derived from the first-order approximation
     and has a certain approximation.

(2) Generally, the system equations of the direct method are all nonlinear equations,
     so the nonlinear filtering method must be adopted, while the system equations of
     the indirect method are all linear equations, and mature linear filtering methods
     can be used.

(3) Each state variable of the indirect method is the error quantity, and the corre-
     sponding order of magnitude is similar. Some of the direct method states are
     the navigation parameters themselves, such as speed and position, and some are
     very small errors, such as attitude error angle. These values vary widely, which
     makes the numerical calculation difficult and affects the accuracy of these error
     estimations.

(4) The direct method can directly reflect the dynamic process of the system, but
     there are still many difficulties in practical application. Only in the inertial flight
     phase of space navigation or in ships with slowly changing acceleration will the
     direct method be used for the system state filter. For an integrated navigation
     system without an INS, if the velocity equation is not needed, the direct method
     can also be used.

Output Correction and Feedback Correction

There are two ways to estimate the system state with the integrated navigation filter: one is to use the estimated errors of each navigation system to calibrate the corresponding output of navigation parameters to obtain the optimal estimation of navigation parameters. This method is called the open-loop method, also known as output correction. The other is to use the estimation of navigation system error to correct the corresponding navigation parameter in the control arrangement of the navigation system. The error estimation is fed back to the interior of each navigation system, and the corresponding error in the navigation system is corrected. This method is called the closed-loop method, also known as feedback correction.

Output correction and feedback correction have their own characteristics:

(1) If the system equation and the measurement equation can correctly reflect the system itself, the output correction and feedback correction are essentially the same, and their effects of estimation and correction are also the same.

(2) The error state in output correction is uncorrected, and the error state in feedback correction has been corrected, so the feedback correction can reflect the true state and dynamic process of system error more closely. In general, for output correction, more complex model system equations should be used to achieve the same accuracy as feedback correction.

(3) Each navigation subsystem in the output correction mode works independently and does not affect each other, so the reliability of the system is high. Feedback correction belongs to depth combination. If one navigation subsystem does not work properly, it will affect other navigation subsystems, so the reliability is worse than output correction.

Common Integrated Navigation Algorithms

(1) Linear Kalman Filter

In the development of Kalman filter theory, the linear Kalman filter (LKF) was first proposed, which was mostly used in indirect integrated navigation systems. Most of the navigation system error equations used in this system are linear equations derived by approximation. The advantage of using the LKF is that the magnitude of each error state is similar, the calculation is relatively small, and it is easy to convert between a single system and integrated navigation. Therefore, the LKF is widely used in research on integrated navigation algorithms [31].

(2) Federal Kalman Filter

In 1988, Carlson proposed the Federal Kalman Filter (FKF) theory for designing a fault-tolerant integrated navigation system. It uses the upper threshold of variance to eliminate the correlation of subfilters so that the main filter can fuse the results of subfilters with a simple algorithm. Its uniqueness lies in the principle of information distribution, that is, the outputs of the filter (including state estimation, estimation error covariance matrix and system noise matrix) are appropriately distributed in several subfilters. Each subfilter completes local estimation with its own observation information, and the result is fed back into the main filter to obtain the global optimal estimation. The federated filter designed in practice is globally suboptimal, but its design is flexible, its computational complexity is small, and its fault tolerance is good. For important vehicles with high autonomy requirements, the reliability of the navigation system is more important than accuracy. This algorithm has been used by the US Air Force as the standard Kalman filter algorithm for fault-tolerant navigation systems.

(3) Adaptive Kalman Filter

When the actual estimation error of the Kalman filter is many times larger than the theoretical prediction error, the filter diverges. Even if the Kalman filter is proven to be asymptotically stable in theory, the convergence of the filter algorithm is not guaranteed in practice. The correct application of the Kalman filter depends on an accurate system model and noise error statistical model. Even in the case of poor initial conditions, ideal state estimation results can be obtained in a relatively short time. However, it is usually impossible to obtain accurate models. Many parameters change slowly in the process of system operation, and the errors of various statistical models will lead to greater filter errors. The divergence is caused by model errors and computational errors in the recursive process. The adaptive Kalman filter (AKF) makes the algorithm more accurately adapt to the poor accuracy and dynamic change of the model. The main adaptive algorithms include the attenuation filtering algorithm, limited memory algorithm, multimodel adaptive estimator, innovation variance modulation algorithm and Sage-Husa adaptive filtering algorithm.

(4) Extended Kalman Filter

The Kalman filter originally proposed is only applicable to linear systems, but the actual navigation system is always a nonlinear system. To find a recursive filtering method similar to the LKF, the Extended Kalman Filter (EKF) takes an approximate method to linearize the nonlinear system model [32]. The common method is to expand the function of the nonlinear system by the first order of a Taylor series to approximate the nonlinear system by establishing the linear disturbance differential equation. The linearization method is the key to applying Kalman filter theory to nonlinear systems.

(5) Unscented Kalman Filter

The EKF simply linearizes all the nonlinear models and then uses the linear Kalman filter method to obtain the first-order approximation of the optimal estimation. Its obvious drawbacks are as follows: first, linearization may produce extremely unstable filtering; second, the EKF needs to calculate Jacobian matrix derivatives, which is not an easy task in most cases. In recent years, the unscented Kalman filter (UKF) has attracted much attention as a new extension of the Kalman filter. Unlike the generalized Kalman filter, the UKF filter uses a deterministic sampling method to solve the propagation of high random variables in nonlinear equations. Representation of Gauss random variables (GRV) by selecting the sets of designed points σ, the mean and variance of GRV can be accurately obtained from the weighting of point σ. When GRV propagates in a nonlinear function, the third-order accuracy of GRV mean and variance can be obtained, while EKF can only achieve first-order accuracy. Therefore, it is better than the EKF in approaching the nonlinear characteristics of the state equation and improving the estimation accuracy. This method can be applied to research on reentry vehicle tracking, inertial navigation initial alignment and Kalman filtering in state estimation.

(6) Particle filter

The Kalman filter is an optimal choice when the system satisfies the prerequisite of a linear and Gaussian distribution. However, these conditions are generally difficult to meet in practice. In nonlinear systems, the EKF and UKF still need to satisfy the Gauss distribution of system noise. For non-Gaussian systems, in the late 1950s, Hammersley et al. proposed the sequential importance sampling (SIS) method. Gordon et al. proposed a bootstrap nonlinear filtering method based on SIS in 1993, which laid the foundation of the particle filter (PF) algorithm.

The so-called PF samples a certain number of discrete samples (particles) from a suitable probability density function and takes the probability density (or probability) of the sample points as the corresponding weightings. With these samples and the corresponding weightings, the posterior probability density can be approximately estimated to realize the state estimation. The larger the probability density is, the larger the corresponding weightings of particles are. When the number of samples is large enough, the discrete particle estimation method can approximate the posterior probability density of an arbitrary distribution (Gauss or non-Gauss) with high accuracy, so the method is not limited by the posterior probability distribution.

(7) Fuzzy Control and Neural Network

Neither the fuzzy logic method nor the neural network method need to establish accurate mathematical models (such as transfer functions and state equations), which are estimated by sample data or prior experience. If all input–output relations of the system, such as transformation, mapping, rules, and estimation, are considered as mathematical models, then both will establish a generalized mathematical model of the system. At the same time, they have strong fault tolerance. Deleting a neuron from a neural network or deleting a rule from a fuzzy rule will not destroy the performance of the whole system. Both of them belong to nonlinear control and have similar complementary structures. In the 1990s, the theory of adaptive-network-based fuzzy inference system (ANFIS) fusion control based on neural network fuzzy inference gradually attracted people's attention. Researchers began to attempt to adopt ANFIS theory to address the divergence problem caused by inaccurate models in the filtering process. In the late 1990s, this method began to be applied to positioning system research. The value of the above methods lies in finding a combination of artificial intelligence (AI) theory and traditional Kalman filter optimal estimation theory to give full play to their respective strengths. It has a good effect on solving obvious model divergence and can also play a role in improving the accuracy of the system. ANFIS Kalman filter technology will become a research hotspot and provide a new intelligent algorithm solution for solving problems in integrated navigation systems.

All Sources Positioning and Navigation

Since 2010, All Sources Positioning and Navigation (ASPN) has become a research focus of public concern of the US military. At the beginning of the research, the focus was to effectively solve the excessive dependence of military weapon systems on GPS. This technology can be traced back to the research on All Sources Adaptive Fusion (ASAF) technology and Signal of Opportunity (SoOp) technology published by the Royal Air Force in 2007. Later, in 2010, the Defense Advanced Research Project Agency (DARPA) officially listed it as one of the five core technologies for comprehensively improving battlefield PNT capabilities. Subsequently, relevant research information was published successively, and in 2017, Draper Laboratory, Vispieres and other organizations jointly approached key problems and completed the research work of relevant information hardware and software architecture, fusion theory, algorithm system, demonstration system, test verification, etc., in three stages. The new information fusion architecture completely solves the problem that the traditional integrated navigation system does not allow dynamic changes in sensor availability. It can increase or change the type and number of navigation sensors at any time when the system is working and can also adaptively switch the fusion algorithm according to the changes in platform motion conditions and environment. This all-source adaptive navigation method can use all available information in real time and calculate accurate navigation results according to the dynamic changes of the carrier. The biggest difference from traditional methods is that ASPN not only needs to solve the problem of multisensor fusion but also needs to provide a unified sensor information interface and a unified navigation filtering algorithm to adapt to the dynamic changes in available navigation information sources.

In addition, researchers such as London University in the United Kingdom also proposed modular solutions to some major problems existing in current multisensor integrated navigation. The British Defense Briefing reported in July 2017 that the US Air Force Research Laboratory successfully developed an ASPN device, which can effectively help fighter aircraft navigate when there is no GPS signal. According to the report, ASPN devices have started to be used in the army and navy. In the future, ASPN devices may become a substitute for the integrated navigation system.

9.3   Integrated Navigation System

9.3.1   The Concept of Integrated Navigation System

With the development of integrated navigation technology, people's understanding of integrated navigation is constantly changing. Integrated navigation systems are becoming more complex, the connotation of navigation technology is becoming richer, and various systems related to integrated navigation are constantly emerging and rapidly developing, such as integrated navigation systems, integrated bridge systems (IBSs), intelligent ships, unmanned aircraft vehicles (UAVs) and other related technologies.

For the narrow sense concept, the information combination part of the navigation system is called integrated navigation equipment, which is different from the main navigation equipment, such as INS and GNSS. In fact, it refers specifically to the information center and the related auxiliary component of integrated navigation.

For the broad sense concept, a generalized integrated navigation system is referred to as an overall system, including all navigation subsystems (such as INS, GPS, log, gyrocompass, deep sounder, meteorological fax machine, anemometer and wind direction meter, etc.) and integrated navigation equipment (information center, integrated navigation display control station, track instrument, electronic chart, etc.) that are responsible for providing navigation information and related environment information for ships.

9.3.2   NAVSSI Ship Integrated Navigation Information System

The NAVigation Sensor System Interface (NAVSSI) of the ship integrated navigation system is a widely used positioning, navigation and timing scheme for surface ships of the United States Navy, which realizes the unification of PNT data in ship formation and provides an important basic information guarantee for formation navigation. It is the core component of the U.S. shipborne navigation system. By collecting different shipborne navigation sensor information and comprehensive processing, it can provide high reliability, high precision position, attitude, speed, time and other information and realize real-time distribution to meet the needs of ship navigation and combat. Initiated by the US Navy in February 1991, NAVSSIs have experienced five stages of development from Block0 to Block4.

NAVSSI Basic Architecture

The NAVSSI employs an open architecture design, and its basic composition is as follows:

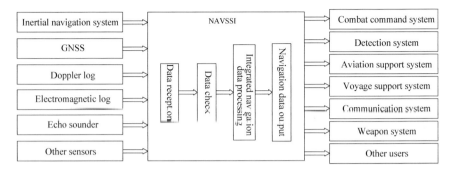

Fig. 9.5 NAVSSI data flow chart

(1) Real-Time Subsystem (RTS): responsible for collecting, processing and distributing PNT data, using the Navigation Data Source Integration (NSI) algorithm to process the data fusion of each navigation system.

(2) DCS: DCS communicates with RTS through a local area network, provides a human–machine interface, ECDIS and ARPA radar interface for RTS, displays navigation information in real time and controls the working state of the RTS.

(3) Network Remote Station (NRS): provides a remote terminal for people in the bridge or other cabin to display and control.

The NAVSSI data flow chart is shown in Fig. 9.5, which mainly includes receiving navigation data sources, data checking, data processing, data output and so on.

The main navigation data sources include INS, Doppler logs, satellite navigation systems, deep sounders, gyrocans and anemometers.

NAVSSI Information Fusion Processing

NAVSSI monitors the data of each navigation information source online and tests the integrity to ensure that the accessed information is in good condition. Once it is detected that the data error continues to exceed a certain threshold, the data are marked invalid. The NSI algorithm isolates the data and gives warnings. The NSI algorithm integrates the GNSS and INS data, fully considers their error characteristics and provides a high-precision, strong robustness position, speed solution scheme to meet the ship's requirements for navigation.

The RTS updates the position estimation based on the obtained optimal speed data, integrates the position data of multiple GPS receivers and INS with the highest precision, calculates the speed by filtering, and corrects the speed data of INS with the estimated speed error. NAVSSI maintains INS accuracy when only INS data are available.

The position data are obtained by comprehensive calculation of GPS and INS data. The optimal INS data are used to calculate the data by interpolation. This method can provide a position with an accuracy of 12 m under normal working conditions for users, which is much higher than that of INS.

The attitude and angular velocity are taken directly from the optimal INS and are not preprocessed. If INS is wrong, its angle value will be marked invalid. A polynomial match will be made to the new heading and attitude data that RTS received. This matching will be used to calculate the attitude data and their change rate according to the latest INS data and time. By using this matching and extrapolation algorithm, the accuracy of the NAVSSI attitude can be guaranteed.

Based on the estimation of the error characteristics of available sensors, the navigation information fusion algorithm ensures the accuracy of the NAVSSI output and distributes the data according to the needs of the user. For users with different accuracy requirements, the effective number of bits of different data can be set.

Since GPS is vulnerable to interference and incapable of meeting the demand of ships for reliable and accurate time, NAVSSI provides a precise timing unit (PTU), which includes a frequency demodulation module (FDM), a rubidium oscillator, and a buffer output device. As soon as the estimated time error exceeds 100 ns, the PTU will turn to work and maintain the time accuracy.

9.4 Matching Integrated Navigation

Modern navigation technology must make full use of the characteristics of the environmental physical field. There are different kinds of physical fields; some are relatively stable, and some change greatly. The physical field features include physical field vectors, isolines, images and patterns. In essence, navigation technology uses various measurement information to calculate navigation parameters. The direct positioning introduced in the previous chapters actually measures the relevant information between various artificial targets or natural targets and vehicles by means of radio, underwater acoustic, optical and other technical means. In fact, they are also measured by various physical quantities or physical field characteristics. Inertial navigation technology is essentially also used to measure the local gravity vector, the earth rotation vector and so on to achieve navigation solutions. Therefore, physical field characteristics are a common feature of navigation technology.

In addition to direct positioning and dead reckoning, there is another method based on environmental features, matching positioning. In fact, navigation based on environmental feature matching has a broader and general form. For example, people and animals will naturally use environmental features and compare them with maps, photographs, text descriptions or memories to determine their location; the location of the user can be inferred directly from the scene image of the environment; or can be determined according to the depth of water or the height of the terrain; or can be achieved according to some conditions such as the vehicle only travels on the road or track, or the pedestrian route does not cross the wall. Integrated positioning methods using different types of environmental characteristics, such as earth magnetic field anomalies, gravity anomalies, and pulsars, have been widely used in many fields.

Environmental feature matching navigation is often used in combination with INSs. Most environmental feature matching navigation systems cannot work independently and are always auxiliary navigation systems that rely on other navigation systems. Environmental feature matching navigation systems are also not global navigation systems and can only work in specific matching areas. Therefore, in essence, environmental feature matching navigation is a special kind of integrated navigation system based on an environmental information database, so it can also sometimes be called a matching integrated navigation system.

With the development of modern measurement and control technology, computer technology, display technology with high resolution, and image processing technology, such as satellite navigation and inertial navigation, matching navigation has become an important technology today.

9.4.1 Basic Principles of Feature Matching Navigation

According to the working principle, feature matching navigation can be divided into two categories: reference map matching navigation and scene matching navigation.

Reference Map Matching Navigation

Different from direct positioning, map matching navigation is a relatively complex positioning method with complex mathematical principles. It is based on a geographic information database that is established on a large number of accurate measurements and mathematical models and achieves positioning according to the real-time observation of geographic information and vehicle motion information. Unlike the positioning method based on angle or ranging measurement, it is more likely to obtain the topographic height of the local environment or physical field data such as magnetic field, gravity, and electric field, rather than the spatial geometric relationship between the vehicle and known points.

The matching navigation was first applied to the terrain reference navigation system and then extended to gravity matching navigation and geomagnetic matching navigation. A variety of matching navigation algorithms based on datum maps have appeared in the fields of geomagnetism, topography, gravity and so on, mainly TERCOM, ICCP and some improved algorithms. Although the algorithm form is different, the main principle is similar. In essence, matching navigation technology is a data association technology.

The principle of the matching navigation system is as follows:

Let the association algorithm be $D(X, Y)$, where X is the track on the datum map and Y is the measured track. If there are nc tracks to be selected in the search domain of geomagnetic, terrain, gravity and other matching maps, the set is formed as follows:

$$C = \left\{ X_j | j = 1, 2, \cdots, n_c \right\} \qquad (9.16)$$

The set P of positions corresponding to nc tracks in set C to be detected

$$p = \left\{ P_j | j = 1, 2, \cdots, n_c \right\} \qquad (9.17)$$

where X_j corresponds to P_j.

Ideally, there must be a point in P that is closest to the real position as the best matching point P_b; the position detected by the association algorithm is called the matching position Pm, which satisfies

$$m = \arg\left\{ \max_{j} \left(D(X_j, Y) \right) \right\} \quad j = 1, 2 \ldots, n_c \qquad (9.18)$$

If P_b is consistent with P_m, it is called the correct interception. Correct interception is the main target of matching algorithm design.

Scene Matching Navigation

Scene matching is a method of matching by directly acquiring images, also known as surface two-dimensional image correlation. Scene matching matches the scene in the moving area of the carrier with the prestored digital scene in the relevant area of the computer through the digital scene matching area correlator to obtain high-precision navigation information, which is often used for terminal guidance [33].

Digital scenes can be remotely obtained by TV cameras, SAR and other image processing devices. Correlation matching is mainly based on the position matching between the remote sensing scene and the computer stored reference image, calculating the correlation amplitude, and comparing the results with the correlation judgment threshold. If the correlation amplitude is higher than the threshold, it indicates effective correlation. By repeating the above related process, the position deviation of the carrier traveling to the area indicated by a reference map can be calculated, and then the carrier navigation route can be corrected.

Figure 9.6 shows the matching navigation positioning system based on underwater geomorphic images. The system obtains the measured geomorphic image through the side scan sonar system or the multibeam system and matches it with the background seabed geomorphic image. According to the corresponding relationship between the two sets of image pixels after matching, the system obtains the position of each pixel of the measured geomorphic image from the background seabed geomorphic image and then obtains the current position of the submersible.

Scene matching technology belongs to the field of image recognition. Common methods include image matching methods based on object boundary lines, Chamfer image matching methods, automatic image registration methods based on differential correction, Scale Invariant Feature Transform (SIFT), etc.

Fig. 9.6 Schematic diagram
of the Shark-S450D side
scan sonar system

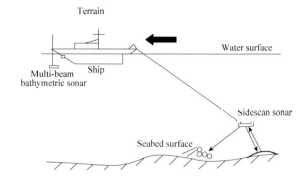

Specification and Characteristics of Matching Navigation

(1) Specification of Matching Algorithm

The performance of the matching algorithm directly affects the accuracy and
reliability of the matching navigation system. Its main specifications are as follows:

  a. Matching probability: This refers to the probability that the true position can
      be correctly detected within a certain precision range. The higher the matching
      probability is, the more reliable the algorithm matching results are.
  b. Matching accuracy: This is an important specification for evaluating the perfor-
      mance of the matching algorithm, which is also the basis for reasonably defining
      the matching probability. Only the method that meets the precision requirement
      can meet the design requirement.
  c. Matching speed: This shows the efficiency of the matching process.
  d. Adaptability of the algorithm: This shows the adaptability of the algorithm
      to the limitations of the principles, calculation conditions and application
      environment.

(2) Characteristics of Matching Algorithm

  a. Limited by the accuracy of the reference database

The establishment of the reference database is mainly obtained by combining the
model calculation with the measured data, and most of the modeling calculation is
completed before navigation. The models adopted include the terrain shape model,
Earth gravity model, and Earth geomagnetic model. According to the different precision requirements, the model order selected is also quite different. The higher the
accuracy of the model is, the greater the computation needed. At the same time, to
further improve the accuracy of the data in the concerned area, modification by the
measured data is also adopted.

  b. Limited by the accuracy of the measurement

Matching navigation must rely on the measurement of environmental information
such as topography, gravity, geomagnetic field, etc. The measurement accuracy is
mainly disturbed by the sensor accuracy and the environment. Sensor accuracy

mainly involves the measurement error of radar altimeters, gravimeters, magnetometers and so on. The disturbance from the environment includes, for example, geomagnetic measurements disturbed by the vehicle magnetic field, underwater gravity measurements affected by gravity changes at different depths, gravity gradient measurements disturbed by environmental quality, and topographic measurements affected by changes in water depth with tides.

c. Area selection and path planning

The accuracy and reliability of matching navigation in the feature area are high. When the feature is not obvious, the navigation reliability will be reduced, and false matching will easily appear. Therefore, we need to set up a special matching area and make the matching routine planning of the vehicle.

9.4.2 Main Matching Navigation Systems

Although there are many kinds of matching navigation systems, their working principles are similar. The matching navigation systems mostly consist of four parts: matching system, INS, digital map storage device and data processing device.

The basic process is similar: first, according to the track obtained, characteristic data of the background datum map (topographic map, gravity map, geomagnetic map, etc.) from the database and then through measurement to obtain the actual measurement data. Then, both are combined to obtain matching locations and other navigation data.

Terrain Referenced Navigation

Terrain reference navigation (TRN) is the earliest application of matching navigation technology, which uses geometric fluctuations of terrain to achieve autonomous navigation. The technology can be traced back to the 1950s, when aerial navigation based on radar altimeters was developed, reaching an accuracy of up to 50 m. TRN can meet the navigation and terrain collision avoidance requirements of tactical missile and aircraft maneuver flight control and has an obvious effect on many tactical flights, such as hedgehopper, strong strike, penetration, interception, etc. Submarine TRN can improve the stealth of submarine navigation and combat capability. The key technology of TRN systems includes digital maps, storage technology, terrain random linearization technology, multimode Kalman filter technology, digital correlators and so on. In addition to radar TRN systems, there are also many TRN systems, such as sonar TRN and laser TRN systems.

The basic principle of the underwater TRN system is as follows (Fig. 9.7). First, the underwater unmanned vehicle (UUV) is allowed to pass through the area with some topographic characteristics. Using a multibeam sonar echo sounder, the UUV can measure the distance of the vehicle to multiple points of the seabed and obtain

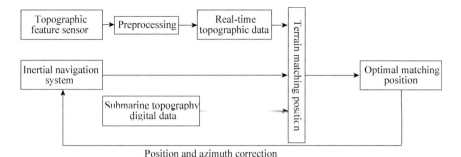

Fig. 9.7 Basic principles of underwater TRN system

the relative position between these points. These points form a bathymetric profile and are then matched with the background terrain database stored in the computer by batch processing to obtain the current position of the UUV. The output of the INS is corrected to limit the accumulation of INS errors and realize the accurate navigation and positioning of the submarine. Common algorithms include the matching algorithm based on the direction, image, chain code of the fathom line, or shape feature, etc.

Similar to other forms of TRN, the accuracy of the system depends on the terrain height change and sensor resolution. The sensor can measure more than 10,000 points at the highest resolution at the same time, and the positioning accuracy can reach 1 m, but the general accuracy is approximately 10 m. If the database is not available, we can calculate the position by comparing the continuously overlapping bathymetric profiles to obtain the velocity in the body coordinate system.

Fig. 9.8 KSS31M
gravimeter, with permission
from reference [first institute
of oceanography, MNR]

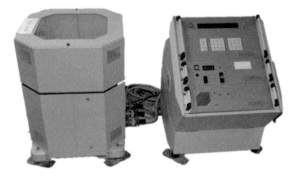

Fig. 9.9 DGS gravimeter, with permission from reference [first institute of oceanography, MNR]

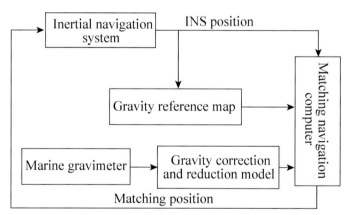

Fig. 9.10 Block diagram of the basic principle of gravity matching-aided navigation

Gravity Aided Navigation

(1) Gravity Aided Navigation

The gravity-aided INS consists of INS, gravity database, gravimeter (Figs. 9.8 and 9.9) and matching navigation computer (matching algorithm embedded). Gravity-aided navigation occurs in two ways. One is to use the echo sounder and the gravimeter installed on the UUV as the measuring equipment. According to the gravity anomaly data obtained in real time and matched with the gravity anomaly map stored on the computer, the EKF is used to estimate the vehicle position in real time. The other is to use the gravity gradient data obtained by the gravity gradiome-ters on the UUV in real time, to match the gravity gradient map stored in the vehicle computer, and to use the EKF to estimate the various navigation errors. In the system with gravimeter and gravity gradiometers, not only can direct matching by gravity anomaly or gravity gradient be used, but the estimation of velocity error in the Kalman filter can also be carried out by using the gravity gradient to calculate the gravity anomaly and vertical deviation in real time. The INS system is to be corrected based on the above information, the INS positioning errors are compensated, and the period of recalibration is prolonged (see Fig. 9.10).

(2) **Key Technologies of Gravity Navigation**

Gravity-aided navigation systems are known as next-generation underwater navi-gation systems. Because there is no energy radiation when obtaining gravity infor-mation, it has good concealment and can correct INS under water. Therefore, it can realize high-precision passive navigation. In addition to the development of high-precision gravity sensors, there are several key technologies:

a.   Real-time gravity measurement system based on moving base

The current marine gravity sensor cannot be directly used for real-time gravity measurement. It needs to add filtering, altimetry, stabilization module and various compensation or correction modules. Some even need to be specially designed and developed for the moving base, such as gravity gradiometers for gravity-aided navigation in the Bell laboratory.

b.   High precision gravity map and data processing theory

Any navigation system based on a gravity field map must establish an accurate datum gravity database in advance; otherwise, it cannot work effectively. At present, the main methods of gravity field measurement are satellite height inversion, aerial gravity measurement and ground point measurement. The first two methods have high measurement efficiency and fast speed but low resolution and precision. The last method has low efficiency and high cost, but high resolution and precision can be obtained. Because of the special characteristics of ocean geography, marine gravity field data obtained in a single way cannot meet the requirements of high-precision navigation over a wide area. Therefore, it is necessary to use all kinds of gravity measurement methods and various mathematical statistical theories to obtain the digital gravity field map needed for navigation.

c.   Gravity matching theory and algorithm

Gravity matching theory and algorithms are the core of underwater gravity matching-aided navigation. The performance of the matching algorithm directly affects the positioning accuracy and reliability of the matching-aided navigation system. At present, most of the studies on gravity matching algorithms adopt widely used terrain matching algorithms, such as sequence iterative matching algorithms and single point iterative matching algorithms. From the international research results, the method has been successfully applied.

Geomagnetic Aided Navigation

Modern geomagnetic navigation technology is based on the geomagnetic field, which is a vector field, and its magnitude and direction are functions of position [34]. At the same time, the geomagnetic field has the characteristics of total intensity, vector intensity, magnetic inclination angle, magnetic deflection angle and intensity gradient, which provides a variety of information for geomagnetic matching. Therefore, the geomagnetic field can be regarded as a natural coordinate system, and the measurement information of the geomagnetic field can be used to realize the navigation and positioning of aircraft, ships, underwater vehicles and so on. The change in the magnetic field in near-Earth space is mainly the change in the geomagnetic anomalous field. The change in the magnetic field in the near-Earth orbit is mainly a change in the main magnetic field, and the change in the geomagnetic anomalous field is obvious. Therefore, the matching method can be used to realize the geomagnetic navigation and positioning of near-Earth space vehicles [35].

The basic work flow of geomagnetic-aided navigation is as follows: First, the geomagnetic data of the preselected area are made into a datum map, and the geomagnetic database is constructed. When the vehicle passes through the corresponding area, the geomagnetic sensor measures the geomagnetic field intensity in real time. The measurement is converted into the required feature quantity through data processing, and the real-time map is constructed and input to the matching module. On the basis of the indicated position, the real-time map and the datum map are matched to determine the matching position of the vehicle. Finally, the matching position is fed back to the integrated navigation filter. The whole system works in closed-loop mode, and the ultimate goal is to compensate for the system error and achieve high precision navigation. The module block diagram of the system is shown in Fig. 9.11. The geomagnetic matching navigation algorithm is the core of the geomagnetic matching system and is commonly using TERCOM and ICCP algorithms.

At present, many key technologies for underwater geomagnetic navigation are still in the exploration stage, but the application of high-precision magnetic sensors, as well as breakthroughs in geomagnetic interference modeling, magnetic sensor configuration detection, geomagnetic map continuation, geomagnetic matching algorithms, integrated navigation theory and so on, will greatly promote breakthroughs in key technologies of underwater geomagnetic navigation systems and promote the development and application of geomagnetic navigation technology.

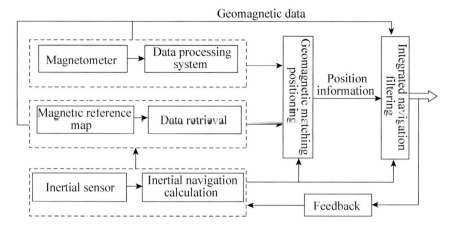

Fig. 9.11 Block diagram of the geomagnetic aided navigation system

Appendix 9.1: Questions

1. How can integrated navigation be understood as essentially a parameter estimation system for multisensor fusion? Explain the relationship between integrated navigation and information fusion.
2. Please briefly describe the different INS-based integrated navigation methods, features, and applications?
3. Please briefly describe the different GNSS-based integrated navigation methods, features, and applications?
4. As an important theoretical basis for integrated navigation, Please briefly describe the basic principles of the linear Kalman filter, and specify its preconditions.
5. Can you understand the recurrence Bayesian estimation? What are the similarities and differences between recursive Bayes estimation and Kalman filtering? Why can the Kalman filter be unified within the recursive Bayesian estimation framework?
6. Please briefly describe the types, development and characteristics of modern integrated navigation information fusion technology?
7. Summarize the basic methods, main types, application fields and characteristics of matching navigation, and explain why matching navigation systems often need to be combined with other navigation systems such as INS.
8. Combined with terrain matching, gravity matching and geomagnetic matching, please explain the composition of the matching navigation system and the key problems.

Appendix 9.2: The Professional Application of Holistic Thinking

Holistic thinking is to further understand and analyze things as a whole on the basis of comprehensive analysis, and thus enables you to grasp the essence what is investigated. The holistic thinking is in fact determined by the objective integrity of things: Just as the parable of blind men and an elephant implies, we can only know the actual appearance of the elephant based on an overall understanding of different parts. Sometimes the acquired information could also be incomplete, and thus holistic thinking, as a higher level of cognition, enables people to grasp the fundamental essence.

Take navigation as an example. The essential purpose of navigation is to accurately guide the carrier movement in four-dimensional space and time. Since people live in the four-dimensional space-time frame, there is always the need for navigation. The changes from ancient to modern times often lie in the scope of navigation, the quality of the environment, the degree of accuracy, the duration of time, the amount of information, the speed of decision-making and so on. The method to determine its own space-time benchmark must be the unification of independent systems and external aids. Autonomous inertial navigation and time-keeping system inevitably require a smaller sensor, while the external sensors detect the space-temporal relationship with the external reference through heat, force, sound, light, electricity, magnetism and other ways to obtain the spatiotemporal reference under the coordinate system. Neither one can do without the other, and both should be obtained and integrated to achieve a more accurate, faster, and more stable overall performance.

As a matter of fact, when things are composed of several parts, they can be at the same time part of a higher-level system. Therefore, the complexity of things is not only associated with the relationship between their internal parts but also comes from their external context. In addition, in the process of problem solving, we should take into consideration both the connection between the problem with others through horizontal comparison and its longitudinal changes. The truth is that the horizontal and the longitudinal perspectives are never independent forms of thinking but rather as an organic whole. Thinking holistically means to take the extensive scientific knowledge as the reference and then to analyze things from a multi-dimensional (e.g., horizontal, longitudinal, inside, and outside) perspective within the space-time setting. In short, we should not confine ourselves to three-dimensional thinking, but, instead, adopt four (or even higher) dimensional thinking.

Chapter 10
Integrated Bridge System

Thanks to the development of computers, modern control, information processing, communication and navigation technology, based on integrated navigation systems combined with ARPA radar, electronic charts, AIS, autopilot and other navigation and ship control equipment, the current integrated bridge system (IBS) performs better functions such as comprehensive navigation, ship control, automatic collision avoidance, integrated information display, communication and navigation management control. It is an integrated system of maritime navigation, communication, radar, control and monitoring. This chapter briefly introduces the integrated bridge system and its related equipment.

10.1 Radio Navigation Aids

Radio navigation aids are mainly used to ensure the safety of ship navigation and to improve the efficiency of ship navigation. The commonly used radio navigation aids are navigation radar, automatic ship identification systems, remote identification and tracking systems and INMARSAT receivers.

10.1.1 Marine Navigation Radar

Marine navigation radar is used to detect the surrounding target position to implement navigation avoidance and self-positioning of the ship (Fig. 10.1). It is the necessary navigation equipment for ship navigation, entry, exit, positioning and narrow channel driving at night.

Radars started to be equipped on ships in the Second World War, and after the war, they began to apply to civilian ships. Embedded with the target plotting function, it

© Science Press 2024
H. Bian et al., *Essentials of Navigation*, https://doi.org/10.1007/978-981-99-5636-4_10

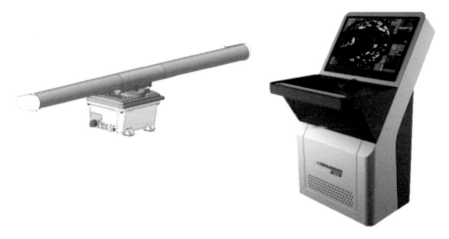

Fig. 10.1 Vessel Atlas Radar, with permission from reference [Beijing Highlander Digital Technology Co., Ltd.]

is thus called automatic radar plotting aid (ARPA). In the late 1960s, an automatic radar target tracking and estimation system was developed that can handle radar video signals, detect and track targets, measure the relative movement between the ship and the target, predict the future movement and closest point of the target, and assist drivers in taking avoidance actions. The general radar represents itself as a fixed point at the center of the radar display. However, because the ship itself moves relative to the fixed target or the moving target, the accurate speed and course of the ship itself must be automatically entered. Ships are usually equipped with one X-band and one S-band radar.

10.1.2   Vessel Traffic Services System

Vessel Traffic Services System (VTS) is implemented by competent authorities to improve ship traffic safety and efficiency and to protect the environment. Within VTS-covered waters, such services should interact with traffic and respond to changing traffic conditions. Initially, it mainly ensured safe and rapid access to ports and through rivers and narrow waterways. In 1948, the first radar used for port surveillance in the world was established using marine radar at the port of Douglas, which solved the problem of water navigation management under the condition of insufficient visibility. Since 1970, a more complex and perfect ship traffic management system with computers as the center has emerged, which is composed of several subsystems.

These subsystems include the very high frequency (VHF) communication subsystem, radar subsystem, integrated radar data processing subsystem, information transmission subsystem, management information subsystem, VHF direction finding subsystem, closed circuit TV monitoring subsystem, information recording subsystem, hydrometeorological subsystem, navigation signal subsystem and amplification broadcasting subsystem. At present, the superiority of VTS is gradually being recognized by people, and it is being built and used more and more widely around the world.

10.1.3 Automatic Ship Identification System

Automatic ship identification system (AIS), which is composed of shore base (base station) facilities and shipboard equipment. It is a digital navigation aid system that integrates network technology, modern communication technology, computer technology and electronic information display technology.

The ship AIS system was developed on the basis of the ship and aircraft IFID. It is closely related to the VTS, which is mainly used in ship management of key waters such as ports but has limitations in operation range and cannot meet the ship information interaction at sea. Born in the 1990s, AIS is a new ship and shore-based broadcasting system working in the VHF frequency band. With the help of GNSS, the AIS transmitter can broadcast ship dynamic information such as carrier position, ship speed, course change rate and heading, combined with ship static data such as ship name, call sign, draft and dangerous cargo, and the VHF channel can broadcast the information to ships and shore stations in nearby waters continuously and automatically without crew intervention. When the surrounding ships are plotted, the nearest encounter point and the recent encounter time of the two ships can be calculated. The shipborne AIS can provide information to avoid collisions quickly, automatically and accurately.

It is of great help to the ship's safety to enable the adjacent ships and shore stations to be updated with the dynamic and static information of all the ships nearby, to communicate and coordinate with each other immediately and to take necessary evasive actions.

10.1.4 Long Range Identification and Tracking System

AIS uses the VHF band and has limitations in that it is difficult to meet the user's demand for remote ship monitoring and management. To solve this problem, the ship long-range identification and tracking system (LRIT) came into being.

The main uses of LRIT are as follows. Firstly promote maritime security, including the security of ships, nearshore and ports; Secondly provides information support for maritime search and rescue, and thirdly provide information support for environmental protection to investigate illegal discharge at sea, oil spill accidents, etc.

10.1.5   INMARSAT Communications Systems

The INMARSAT system is a successful commercial DAMA synchronous satellite communication system that uses a limited 34 MHz bandwidth of frequency resources to provide services to nearly 150,000 communication terminals worldwide.

The INMARSAT communication system consists of three parts: the INMARSAT satellite, ground station and terminals. The onshore station is the ground transit station for satellite communication. The ship station is the maritime user station, set on all kinds of maritime platforms for navigation. The antennas of the ship station are equipped with stable platforms and tracking mechanisms, so that the antennas can always point to the satellite when the ship sways and tilts. According to the requirements, ships transmit communication signals to INMARSAT satellites in the orbit of geostationary satellites, and transmit them to the shore stations through the satellites, which then realize mutual communication with land users around the world through the ground communication network or the international satellite communication network. In addition to being widely used in telephone, telegraph, telex and data transmission services, INMARSAT satellites also have rescue and navigation services. The system transmits data such as heading, speed and position to the shore station at any time. Once the ship has an emergency at sea, the shore station can quickly determine the specific location of the ship and organize rescue in time. The system can also navigate for ships at sea.

10.2   Hydrometeorological Equipment

The hydrometeorological conditions of the sea affect ships all times, which has an obvious influence on ship safety. The marine hydrological data required for ship safety navigation are extensive, including sea currents, wind, surge, waves, sea temperature, salinity, water color, transparency, tidal current, acoustics, sea ice and so on. A hydrological measuring device is an instrument that provides detailed hydrological information for ships. There are many kinds of such devices.

10.2.1   Deep Water Temperature Measurement

The measurement of deep water temperature mainly utilizes traditional reversing thermometers, self-contained conductivity/temperature/depth instruments (such as STD, TD and CTD) (Fig. 10.2), electronic temperature depth instruments (EBT), expendable bathythermograph (XBT) and so on. The seawater temperature of each water layer on the vertical section can be measured directly by these instruments.

Fig. 10.2 Seawater temperature and salinity measurement (CTD) instrument, with permission from reference [First Institute of Oceanography, MNR]

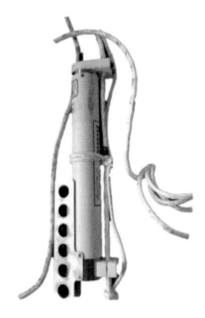

Fig. 10.3 Doppler current profiler, with permission from reference [First Institute of Oceanography, MNR]

10.2.2   *Observation of Seawater Salinity*

(1) **Conductimetric Method**: The conductivity of seawater fluctuates with variations in seawater temperature, pressure and salinity. At the same temperature

and pressure, the conductivity of seawater composed of the same ions is only related to salinity. Accordingly, the salinity of seawater can be measured.

(2) **Optical measurement method**: According to the fact that the different salinities and temperatures of seawater differ in the refractive indices of light, the principle of light refraction is used to determine the seawater salinity. The measuring instruments include the Abbe refractometer, multiprism differential refractometer, and portable refractometer.

(3) **Determination of specific gravity**: According to the international equation of seawater state, the seawater salinity is calculated according to the measured seawater temperature, density, and depth, measured by a gravity meter; however, the precision is not very high.

(4) **Acoustic method**: According to the relationship between sound velocity in water and its salinity, temperature, and pressure, the seawater salinity can be calculated based on the sound velocity, temperature and depth measured by a sound velocimeter, while the accuracy is not high.

Among the abovementioned methods, the conductimetric method is the main method, and other methods can be used as supplementary means in some cases (Fig. 10.2).

10.2.3 Observation of Sound Velocity in Seawater

(1) Indirect sound velocity measurement

Since the sound velocity of seawater is a function of water temperature, salinity and pressure, the sound velocity of seawater can be determined by a specific formula based on the measured water temperature, salinity and pressure data. This method is called indirect sound velocity measurement. At present, there are many empirical formulas describing the relationship between parameters and the sound velocity of seawater, and the Wilson formula is more commonly used in practice. The current temperature, salt and deep measurement devices can basically meet the accuracy requirements of sound velocity calculations.

(2) Direct sound velocity measurement

The method that directly calculates the sound velocity of seawater by measuring the time or phase of the propagation of sound waves at a fixed distance is the direct sound velocity measurement. The sound velocimeter is based on principles such as the time difference method (pulse-echo method), resonant interference method, phase method and sing-around method. At present, the most widely used sound velocimeter are sing-around velocimeter and multiparameter sound velocity profilers.

10.2.4   Current Observations

The main methods of current measurement are mooring, self-propelled buoy and drifting buoy. The shallow sea current is mainly measured by anchorage, while that of the deep sea area is mainly measured by navigation buoys or tracking buoys. The methods of current determination can be divided into indirect methods and direct methods. The indirect technique involves calculating the sea current after determining the density of the water in the ocean based on its temperature and salinity. Generally, it is only applied in the calculation of ocean currents. The direct method is to observe the current directly using various current measuring instruments (such as current meters), which are generally divided into two categories: surface current observation and subsurface current observation. For the currents below the surface, in most cases, short-term point observations are made using current meters such as the propeller type current meter, direct reading current meter, electrical current meter and so on.

10.2.5   Observation of Waves

Wave observation mainly aims at the observation of wind waves and swells. In addition to the wave direction, period, and height, sometimes it is necessary to observe the wave width and its propagation speed. The specific methods of wave observation are mainly to establish an ocean observation station on the shore, select the open sea area with sufficient water depth (more than 1/2 wavelength) and area near the visual field, set up a wave buoy with ropes underwater, and use the stadia wave gauge in the observation station for fixed observation. This method has the advantages of being simple, convenient, economic and practical and it can obtain many kinds of observation data. The disadvantage is that the observation area and observation space are greatly limited, especially when waves in deep water and ocean areas cannot be monitored. In addition, the terrain around the observation station has a great impact on the observation data. At present, there are many kinds of electronic wave gauges for observing waves in different environments, such as shoreside, deep water or ocean, such as WG100 wave gauge produced in the United States. These electronic wave gauges are often limited in the scope of application (including the type of observation elements, the observation sea area, and the water depth).

10.2.6   Measurement of Sea Ice

At present, the technology to monitor and measure the thickness of sea ice mainly includes optical method and upward-view sonar technique. Optical method is visual observations of sea ice. The upward-view sonar technique is the classical method to observe ice thickness, and there are two main forms, namely, submarine sonar profile measurement and upward-view sonar.

10.3   Meteorological Element Observation Equipment

The basic factors that characterize and reflect the physical quantities and phenomena of the atmospheric conditions are called meteorological elements or weather elements, including temperature, air pressure, humidity, wind speed, wind direction, visibility and other physical quantities, as well as meteorological/weather phenomena such as clouds, precipitation and thunderstorms. Meteorological elements change over time and space, and they are the basic items of maritime meteorological observation. Observation data of meteorological elements are the basic data of meteorological support.

The meteorological measurement instruments commonly used by ships include anemometers, meteorometers, weather facsimile receiver, thermometers, hygrometers, barometers, etc. (Fig. 10.4). These meteorological instruments obtain as much as possible meteorological information around the ship by using a variety of different ways and methods.

Fig. 10.4 DZC1 meteorometer, with permission from reference [Aerospace Newsky Technology Co., Ltd.]

Fig. 10.5 WUSH-FVV10
ship visibility, with
permission from reference
[Aerospace Newsky
Technology Co., Ltd.]

10.3.1 Observation of Sea Surface Visibility

Atmospheric visibility can be measured with the unaided eyer by instruments such
as transmissometers and automatic laser visibility sensor (Fig. 10.5). The transmis-
someter is measured directly by the beam through the atmosphere between two fixed
points, and it can then be used to determine the visibility. This method requires the
beam to pass through a sufficiently long atmospheric column, and the reliability of
the measurement is affected by the stability of the light source and other hardware
systems. It is generally only suitable for moderate visibility observations because in
low visibility weather such as rain and fog, due to water vapor absorption in complex
conditions, it will cause major errors.

10.3.2 Wind Observation

Wind observations include wind direction and wind speed, usually measured by
instruments [36]. The ship weather stations and the handheld anemometer are
commonly used wind measuring instruments on ships. A ship weather station consists
of a wind speed sensor, a wind direction sensor, a humidity sensor, and an indicator.
The anemometer can measure wind speed and direction in real time. The common
types are wind cup type and propeller type. In addition, there are hot wire anemometers made on the principle of heat dissipation rate and wind speed correlation of
heated objects, while ultrasonic anemometers are made on the principle that acoustic
wave propagation speed is affected by wind speed. (Figs. 10.6 and 10.7).

10.3.3 Observation of Air Pressure

Air pressure is measured by the aneroid barometer on ships. The aneroid barometer is
made by applying the principle of balance between changes in atmospheric pressure
and the deflection of the flexible-walled evacuated capsule. To get the sea level
pressure at the observation point, the values of air pressure read directly from the
aneroid barometer should be corrected for the index error correction, temperature
correction, supplementary correction, and altitude correction.

Fig. 10.6 Ultrasonic wind
sensors, with permission
from reference [Aerospace
Newsky Technology Co.,
Ltd.]

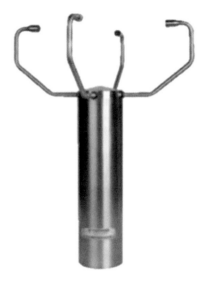

Fig. 10.7 Meteorological
element sensors, with
permission from reference
[Aerospace Newsky
Technology Co., Ltd.]

10.3.4 Air Temperature and Humidity Observations

Psychrometers and ship weather stations can be used to observe air temperature
and humidity on ships. The working principle of humidity measurement of the ship
weather station is the same as that of the psychrometer: by using two identical
thermometers, but the sensing part of the thermometer is different from the dry bulb

and wet bulb of the psychrometer, the ship weather station uses resistance as a sensing element to indicate temperature with a balance bridge circuit.

10.3.5   Weather Facsimile Receiver

A weather facsimile receiver is a fax machine for transmitting meteorological cloud maps and other meteorological charts used for aviation, navigation, etc. It can regularly receive meteorological information, such as weather charts and satellite images sent by the Maritime Safety Administration and the Meteorological Administration, to help the navigator grasp the meteorological information of large sea areas and compensate for the small detection scope of the ship's own meteorological measurement device. Navigators can effectively predict and avoid strong maritime climate phenomena according to cloud maps, providing the necessary guarantee for the safety of navigation.

Meteorological faxes are transmitted in two modes: meteorological wireless facsimile broadcasts via shortwave (3~30 MHz) and point-to-point meteorological transmissions through wired or radio channels. Meteorological fax broadcasting is a one-way transmission mode, and most weather facsimile receivers are only used for receiving information. The size of the transmission is larger than that of a newspaper, but the resolution is not high (Fig. 10.8).

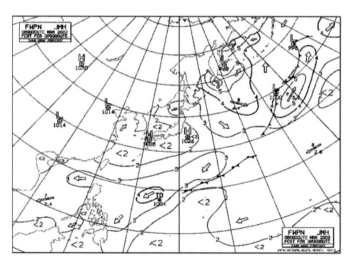

Fig. 10.8   Facsimile weather maps

10.4 Navigation Planning and Recording Equipment

Electronic chart system is a technical general term for all the production, application, software and hardware of electronic chart, showing chart and other navigation information on the display. An electronic chart is "the legitimate equivalent of paper chart". The basics of the chart are briefly described below.

10.4.1 Basic Knowledge of Chart Projection

A chart is a kind of map that is specially drawn for the needs of navigation. The chart is an essential tool for navigation. There are many works such as routes planning, speed calculation, positioning, risk avoidance, the accident responsibility determination must be implemented and completed on the chart.

Projection is an important concept of the chart. The earth ellipsoid is a nonplanable surface, while the chart is a plane, and the way to solve the contradiction between the surface and the plane is chart projection. The chart projection method, scale and orientation control constitute the mathematical law of the chart, which is the basis of chart mapping. This law ensures the charts sufficient mathematical accuracy, measurable and comparable.

Types of Chart Projections

The basic principle of chart projection is as follows: first, project the points on the surface of the earth to the earth ellipsoid vertically, express the points on the earth ellipsoid to the plane according to the mathematical method of chart projection, and finally shrink to the visibility by the scale.

There are many types of chart projections. (1) According to the deformation nature, it can be divided into equal angular projection, equal area projection and arbitrary projection. (2) According to the shape of the latitude and longitude curve projected by the normal axis, it can be divided into azimuth projection, cylindrical projection, and conical projection. (3) According to the relationship between the projection surface and ground axis, it can be divided into normal axis projection, transverse axis projection, oblique axis projection and so on (as shown in Figs. 10.9, 10.10 and 10.11).

The Basic Requirements of Chart

The chart has two basic requirements:

(1) The rhumb line is a straight line on the chart. When a ship is sailing at sea, for the convenience of manipulation, it always maintains a fixed course for a period of time, which is called a rhumb line. The rhumb line is actually a logarithmic

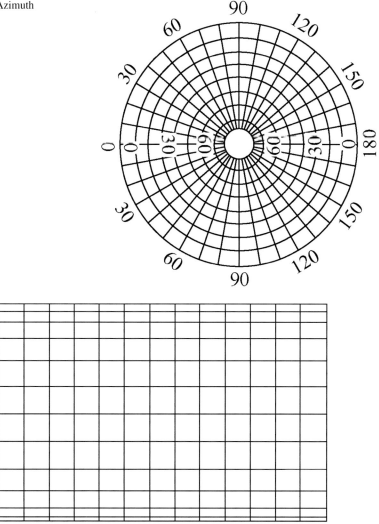

Fig. 10.9 Azimuth projection

Fig. 10.10 Cylindrical projection

spiral curve gradually approaching the earth pole. If the rhumb line is also a curve projected on the chart, it will bring great inconvenience to the plotting on the chart. Therefore, when producing charts, we must satisfy the requirement that the rhumb line is a straight line on the charts.

(2) The property of projection is an equiangular projection. For the convenience of navigation, the angle on the chart should be equal to that on the earth. In this way, the navigation line and the azimuth line can be plotted directly on the chart according to the degree of true course or true orientation.

Therefore, despite the variety of existing projections, there are only three projections in navigation for different applications: the Mercator projection (or equiangular

Fig. 10.11 Cone projection

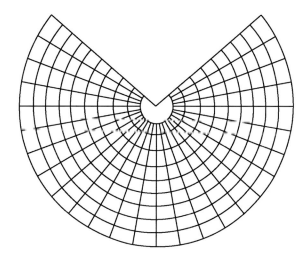

cylinder projection), the sundial projection (or centroid projection) and the Gauss–Kruger projection. Among them, the Mercator map accounts for more than 95% of the current charts and is the most commonly used chart.

10.4.2 Electronic Chart System

The Basic Concept of Electronic Chart System

The electronic chart system (ECS) is an integrated navigation information processing application system that can realize position calculation and integrated processing, display all kinds of navigation data and radar information, and complete the integrated ship navigation and control task [37].

The electronic chart is not a simple copy of a paper chart. After realizing the function equivalent to paper charts, electronic charts further promote the comprehensive realization of ship navigation automation. The essence of navigation automation is the automation of navigation information processing, safe navigation decisions, the establishment of a perfect navigation information processing mechanism, and the realization of the extraction, management, analysis and synthesis of navigation information automation. With the development of electronic charts, their functions as "maps" and "information systems" have become mature and perfect. It can not only be used for chart display, navigation track plotting and provide navigation information but also to analyze and comprehensively process navigation information and related information to serve as an information processing platform to provide underlying support for navigation automation.

Electronic Charts and Their Classification

The chart information displayed by ECS comes from chart survey data, original paper chart digital data and certain information provided by other nautical publications. A chart database was built from these data. The structure and storage of chart databases is a key technology of ECS, which directly affects the required capacity, accuracy and operation speed of electronic chart databases. According to the production method, the electronic chart can be divided into raster and vector charts.

(1)   Raster charts refer to the data information documents made by optical scanning of paper charts, which can be regarded as copies of paper charts and are unable to provide selective queries and displays.
(2)   Vector charts refer to databases that classify and store digitized chart information, which have functions such as selective query, display and data usage, and can also be combined with other systems to provide functions such as automatic alarm in alert areas and dangerous areas.

International Standards and Regulations

Several international organizations have made the following four standards for electronic charts and systems.

(1)   The International Hydrographic Organization (IHO) has developed the Transfer Standard for Digital Hydrographic Data of ECDIS charts (S-57). It describes the standards used for the exchange of digital hydrographic data between national hydrographic services and the data distribution to personnel and ECDIS producers.
(2)   IHO has formulated the Data Protection Scheme (S-63). It describes the organization's recommended standards for the protection of electronic chart information and its appropriate security construction and operating procedures, highlighting the specifications of the corresponding system.
(3)   The International Maritime Organization (IMO) has established ECDIS performance standards. The standard gives the definition of ECDIS and specifies the performance requirements such as information display, chart correction, route design, route monitoring, navigation record, etc.
(4)   The International Electrotechnical Commission (IEC) has developed the ECDIS hardware equipment performance and testing standards (IEC61174). It describes the testing methods and required performance for ECDIS that meet IMO standards. The ECDIS that meets this standard is then certified and legally becomes marine equipment.

System Composition and Function

An ECDIS generally includes three parts: hardware, software and data. At present, there are a variety of ships, different in size, shape and tonnage, ranging from dozens of tons to hundreds of thousands of tons.

A single ECDIS model cannot meet various requirements, so the software and hardware structure must be modular to choose customized development according to the requirements. The shape diagram of ECDIS is shown in Fig. 10.12.

Fig. 10.12 ECDIS, with
permission from reference
[Beijing Highlander Digital
Technology Co., Ltd.]

For the hardware, the ECDIS system should include at least central processing devices, high-resolution display and graphic acceleration cards, ENC carrier, the interface of the central processing device and navigation equipment and radar, etc.

The main functions of ECDIS include chart display, chart operation, chart correction, positioning and navigation information display, navigation information consultation, radar information processing, route monitoring and navigation records (see Fig. 10.13). As a platform for comprehensive navigation automation information services, ECDIS mainly provides navigation environment information based on safe navigation information, provides an operation platform for spatial and attribute data reflecting the characteristics of the marine environment, supports spatial positioning retrieval and analysis, and bridges the interaction between the navigation technology field and other technical fields.

Therefore, the main functions of common ECDIS include chart display, chart operation, chart correction, positioning and navigation, navigation information consultation, radar information processing, navigation route monitoring and navigation records.

Characteristics of Electronic Charts

(1) Improving navigation automation

ECDIS can be combined with navigation systems, AIS, collision avoidance systems and communication systems to form a new type of integrated system, whose function far exceeds the function of paper charts, such as planning routes, route monitoring, automatic chart operations, and automatic record data. It can

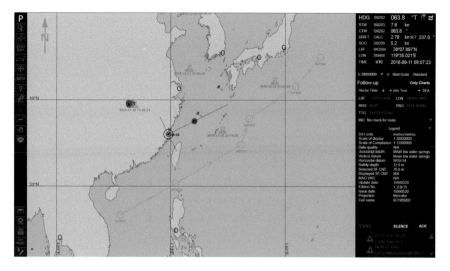

Fig. 10.13 Chart display function, with permission form reference [Beijing Highlander Digital Technology Co., Ltd.]

significantly improve the level of navigation automation and reduce the burden of seafarers.

(2)   Improving the safety of navigation

ECDIS can not only accurately display the position of the ship but also show the hydrogeological characteristics of the planned route, the real-time track and the surrounding environment and can search for potential risk according to the predetermined route. Therefore, it can provide reliable technical support to guarantee the safety and accuracy of navigation. In particular, the navigation radar information will be superimposed on the chart, which can enable the navigator to judge quickly and make the right decision to avoid collision. Navigators can focus on navigation monitoring and automatic alarms, which greatly reduces accidents caused by human error.

(3)   Expanding the chart displaying function

ECDIS can not only display all the information of a paper chart but also display the contents of navigation alarm information, human–machine dialogs, etc. If necessary, additional information on charts, navigational manuals, route guides, etc., may also be displayed. Users can freely choose the displayed sea area, move and roam, zoom in and out, to achieve hierarchical display, or decide whether to display with additional radar information. Color dynamic graphics can improve the recognition of chart target display.

(4)   Electronic chart correction can be realized

ECDIS can easily and efficiently realize chart correction by computer programs automatically.

10.4.3 Track Autoploter

The function of the track autoploter is to draw the ship track automatically in real time on the chart according to the information of ship position, heading and speed provided by the integrated navigation system during ship voyage.

The early version of an autoploter is a mechanical–electrical pen-drawer. To be more specific, the course and speed given by gyrocompass and log are calculated by the electromechanical device (analog calculator) to obtain the ship velocity component along the east–west direction and north–south direction, which are then converted to the relative velocity on the chart by the scale. Such a velocity is used to control the motors in both the X and Y directions, which drive the drawing pen by the deceleration device to draw the track of the ship on the chart. The utilization of electromechanical pen-drawing instruments has improved the quality of chart drawing and minimized mistakes. At the same time, it has released the operators from the trifling work.

At present, the widely used types are electronic display trackers and plane ray projection trackers. They have been greatly improved from the structure to the function and have developed from the single track autoploter into a multifunctional autoploter system.

10.4.4 Voyage Data Recorder

The voyage data recorder (VDR) is commonly referred to as the marine black box (Fig. 10.14). Its main function is to record the data parameters of ship navigation, and it is the special equipment for the investigation and analysis of accidents. To facilitate the search, the VDR is usually in striking orange–red; the shell is solid, and an embedded locator beacon can automatically emit a specific frequency after an accident.

Fig. 10.14 Voyage data recorder, with permission from reference [Beijing Highlander Digital Technology Co., Ltd.]

10.5 Autopilot

Autopilot is an important navigation equipment for ships that is used to automatically adjust the course of a ship to maintain its set route (Fig. 10.15). It can automatically maintain or change the heading of the ship with high precision and thus reduce the speed loss, shorten the navigation time, save fuel and prolong the operational radius of action. Therefore, it is of great significance to long voyage ships.

10.5.1 Basic Concept of Autopilot

The autopilot is conventionally defined as a device used to automatically adjust the heading of ships, and it works based on compass heading, the set heading and the rudder angle. It gives the control signal to control the proper steering gear so that the ship can automatically keep sailing on the set heading. The current autopilot is defined as a device used to automatically adjust the track of ships, and it uses the information of ship motion provided by navigation equipment and gives the control signal through a certain control algorithm. As a result, the steering gear works properly, and the ship sails according to the set navigation route. The differences between the two types are the information received and the control target of the autopilot.

Fig. 10.15 Autopilot, with permission from reference [Beijing Highlander Digital Technology Co., Ltd.]

10.5.2   Basic Principles of Autopilot

The autopilot installed on ships before the 1980s was generally only capable of heading control, which allowed ships to navigate at a given heading. The basic working schematic diagram of the autopilot is shown in Fig. 10.16. As precise GNSS receivers are used on ships, people have begun to work on designing track-control autopilots that can control ships on a given planned route.

The autopilot, together with the integrated navigation system, can constitute an automatic voyage system, with the planned route determined by the integrated navigation system and the precise track controlled by the autopilot. With the combination of ARPA radar, GPS, Loran C, log and other navigation equipment, intelligent integrated control is carried out by computer, and thus, the level of ship navigation automation and navigation safety has been greatly improved. This is the product of the combination of modern control theory and computer technology and a great leap from the heading-controlled autopilot to the track-controlled autopilot.

The basic working schematic diagram of the track-controlled autopilot is shown in Fig. 10.17.

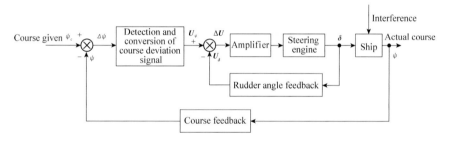

Fig. 10.16 Basic operating principle of heading-controlled autopilot

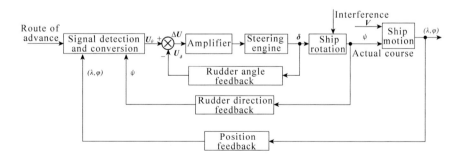

Fig. 10.17 Basic operating principle of the track-controlled autopilot

10.5.3 Control Method of Autopilot

Modern autopilots mainly have three kinds of control methods, namely, P-I-D control, adaptive control and intelligent control.

P-I-D Control

(1) Proportional control (P)

Proportional control takes the ship's heading deviation directly as the modified signal of the steering equipment, and its control equation is:

$$\delta = K_P \psi \qquad (10.1)$$

where δ is the rudder angle signal, ψ is the heading deviation signal, and K_P is the proportional coefficient, which is adapted to fit the load and environmental changes manually and continuously.

The ship's autopilot has the following functions when it is in use: When a ship deviates from heading, $\Delta U = |U_\psi| - |U_\delta| > 0$, and the ship would turn rudder automatically; when a ship returns, $\Delta U = |U_\psi| - |U_\delta| < 0$, and the ship would return rudder automatically; when $\Delta U = |U_\psi| - |U_\delta| = 0$, the rudder stops. K_P should take a smaller value to avoid oscillation.

(2) Differential control (D)

For ships that navigate in static water at low speed, the effect of this proportional (P) control is basically satisfactory, but for ships with complex sea conditions, proportional control is no longer applicable. A more advanced control system should contain derivatives of heading deviations in the form of:

$$\delta = K_p \psi + K_d \frac{\mathrm{d}\psi}{\mathrm{d}t} \qquad (10.2)$$

where K_p is the proportional coefficient and K_d is the differential coefficient rudder angle.

Such control is also called proportional-differential (P-D) control. The control of the rudder angle is not only proportional to the yaw angle but also related to the change rate of the yaw angle. Because the control related to the yaw changing speed is added to the system, the rudder angle can reflect the speed and direction of the yaw change. In other words, when the ship deviates from the planned heading, the differential prediction controller can help to speed up the rudder and increase the rudder angle to prevent the yaw angle from increasing, thus forcing the ship to return. In the return of the ship, a stable rudder angle is generated by the differential predictive controller. To be more specific, a braking torque is generated to prevent the return of the ship from being off more than half of the given heading due to inertia. Adjusting the magnitude of the differential quantity can make the ship quickly stabilize on the given heading.

In conclusion, due to the addition of differential control in the system, the system damping is increased, and the stability margin of the heading control system is increased. As a result, it is possible to select a large open loop gain and to improve the heading accuracy. The realization of differential action is not difficult. When the differential effect is adjusted, the performance of the autopilot can be obviously improved, so most autopilots use differential control.

(3) Integral control (I)

To keep the heading constant, the integral term of heading deviation should be added when there is a disturbance moment of downwind or upper wind caused by transverse wind.

$$\delta = K_p \Delta \psi + K_d \frac{\mathrm{d}\psi}{\mathrm{d}t} + K_i \int_0^t \Delta \psi \, \mathrm{d}t \qquad (10.3)$$

where K_p is the proportional coefficient, K_d is the differential coefficient, and K_i is the integral coefficient.

The magnitude of the rudder angle during this control is related not only to the size and rate of the yaw changing angle but also to the integral of the yaw angle. Figure 10.18 shows the proportional-integral–differential rudder structure control block diagram. Integral control is added to overcome the yaw ψ caused by the constant interference moment and to eliminate the heading steady state error.

Integral control corrects the difference signal so that the ship can still maintain a certain rudder angle when the yaw angle is zero. This produces a rotating torque to counteract the effect of the asymmetric continuous interference moment on the ship's heading. The so-called constant interference torque refers to the draft asymmetry between the ship's left and right sides, the unequal between two propellers, the influence of wind and flow and so on. The common feature of these disturbances is that the direction of the disturbance moment on the ship remains unchanged. It can be divided into two types: one is constant; the other remains the same direction but with constantly changing magnitude. This interference may not always be continuous and can be regarded as a breaking constant interference effect.

The addition of the integral term may reduce the response speed of the rudder and slow the ship's reaction. To counteract this effect, an acceleration term may be added, that is,

$$\delta = K_p \Delta \psi + K_d \frac{\mathrm{d}\psi}{\mathrm{d}t} + K_i \int_0^t \Delta \psi \mathrm{d}t + K_{dd} \frac{\mathrm{d}^2 \psi}{\mathrm{d}t^2} \qquad (10.4)$$

By adjusting these control parameters, such as scale, differential, and integral, we can obtain better manipulation performance. For the interference of the high frequency wave, P-I-D control could be too sensitive. To avoid frequent steering manipulation caused by high frequency interference, the nonlinear weather regulation

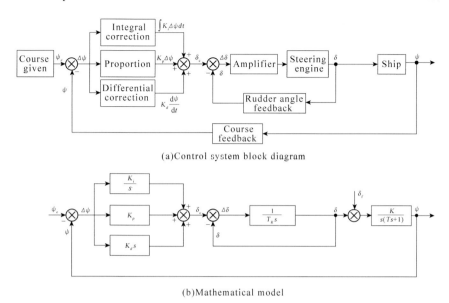

Fig. 10.18 Proportional-integral–differential autopilot structure

of the "dead zone" is often used, but the dead zone will lead to the deterioration of the low frequency characteristics of the control system, which in turn may produce continuous periodic yaw, reduce the accuracy of navigation and increase the energy consumption.

In addition, when the dynamic characteristics of the ship (speed, load, water depth, appearance, etc.) or external conditions (wind, wave, flow, etc.) continued to change, the control parameters needed to be continuously tuned manually. The improper control parameters adopted by the controller will lead to a poor control effect, such as a large steering amplitude and frequent steering manipulation, and manual setting parameters are often troublesome. To solve this problem, adaptive control methods are proposed.

Adaptive Control

The adaptive system can continuously automatically identify (tune) the control parameters of the P-I-D algorithm to suit the dynamic characteristics of ship and environmental conditions. At present, the methods proposed in ship control mainly include the adaptive P-I-D design method, random adaptive method, model reference method, self-correction method based on the conditional cost function, minimum variance self-correction method, linear quadratic Gaussian method, and variable structure method. These adaptive control methods have their own advantages and disadvantages, and adaptive control is still in the process of continuous development.

Intelligent Control

Traditional control methods are very effective for finite-dimensional, linear and time-invariant control processes. However, due to the uncertainty, nonlinearity, instability and complexity of the actual ship system, it is difficult to establish accurate model equations for it, and it cannot be directly represented and analyzed.

Considering that the helmsman can understand and deal with the situation intelligently and control the ship effectively, it is reasonable to seek intelligent control methods similar to manual operation. Although intelligent control has been developed for more than 20 years, it is still at the initial stage. At present, the research on intelligent control mainly covers the basic theory and methods of intelligent control, the structure of intelligent control systems, expert control based on knowledge systems, intelligent control based on fuzzy systems and neural networks, control learning based on information theory and evolution theory (genetic algorithms), intelligent control based on learning and adaptability, and so on.

Due to the advantages and disadvantages of different control methods, the control of autopilots has developed in the direction of integration in recent years to learn from each other in different ways, such as P-I-D with fuzzy control, P-I-D with neural control, fuzzy control with neural control, fuzzy control with genetic algorithm, neural control with genetic algorithm, etc.

10.5.4   Operation Mode of Autopilot

Simple Operation Mode

In this mode, the rotation and stop of the steering blade are both controlled by humans. For example, if the ship needs to turn 5° to the left, the steering gear must be first moved to the turning position toward the left. When the rudder reaches 5° to the left, the steering gear needs to be stopped. This method requires a rudder angle repeater, and to realize heading control, the compass heading and rudder angle repeater must be observed in real time.

Servo Mode

The helmsman gives the steering order, and the steering gear automatically stops under the control of the steering angle following system after turning to the given steering angle. To control the heading of the ship, the helmsman also needs to constantly modify the angle with experience, watching the angle command scale. In addition, the operator needs to keep observing the rudder angle instruction scale and compass to achieve heading control.

Autopilot Mode

In this method, the helmsman only needs to give a certain heading instruction, and the autopilot can automatically adjust the magnitude of the rudder angle under the heading servo system so that the ship can keep moving on the specified heading. The implementation of heading control requires only the input of the instruction heading.

Modes 1 and 2 are nonheading automatic control systems with helmsman as the feedback link, and mode 3 automatically keeps the ship on a given heading after the operator gives a heading command, which is the main working mode of automatic steering.

10.6 Integrated Bridge System

10.6.1 Concept of the Integrated Bridge System

The integrated bridge system (IBS) is a ship automatic navigation system developed from the integrated navigation system in the early 1970s (Fig. 10.19).

For 40 years, with the development of computers, modern control, information processing, communication and navigation technology, based on integrated navigation systems, combined with ARPA radar, ECDIS, AIS, autopilot and other navigation and ship control equipment, IBS can perform various functions, such as navigation, ship control, automatic collision avoidance, integrated information display, communication and control management.

It is an integrated system of maritime navigation, communication, radar, navigation control and monitoring. IBS has played an important role in improving the ship navigation automation, safety and operation efficiency.

To ensure the safety of ships, according to the relevant international regulations, bridges must be equipped with a variety of navigation instruments, such as AIS and

Fig. 10.19 IBS system, with permission from reference [CSSC marine technology]

VDR. Since all the equipment is independent and distributed, there is no effective integration of all kinds of navigation information, which calls for the driver to make a comprehensive judgment according to those information. In fact, it will bring great pressure to the driver in case of emergency and could even cause serious misjudgment.

To reduce the burden of drivers and avoid unnecessary mistakes, the information provided by navigation instruments must be clear, accurate and complete. Therefore, it is necessary to focus on the rational layout of bridge equipment, especially on the integration and optimization of the functions of different navigation equipment. With the rapid increase in the number, tonnage and speed of ships, the safety of ship navigation has set higher requirements, such as requiring ships to sail safely and reliably under global, all-weather conditions; requiring ships to navigate accurately on the most economical routes to reduce fuel consumption; and requiring ships to increase automation to reduce staff and increase efficiency. To meet these requirements, manufacturers and research institutions of navigation instruments began to develop a comprehensive bridge system with multiple functions.

While current IBS is not mandatory installation of equipment under the SOLAS Convention, in recent years, international organizations, such as IMO and IEC, have successively imposed requirements for IBS performance standards, and classification societies have also established requirements for the IBS of different levels of ships.

10.6.2   Development of Integrated Bridges

After over 30 years of development, IBS has experienced four generations. The functions of the IBS have also been developed from the fundamental aspect of information combination to manifold aspects, such as navigation, platform control, ship condition monitoring, equipment management, communication control, intelligent decision making, maintenance diagnosis, and voyage data recorder. The future intelligent ship IBS (I^2BS) would display distinctive features of intelligent system composition, digitalization of electronic information, modernization of testing means, easy operation training and modularization of maintenance.

Systems with Navigation Functions (1960s~1970s)

IBS was first developed in the late 1960s. With the help of IBS, the traditional radar can receive positioning information, and the radar function improves when the planned route and navigation safe haven are input into the radar through the microprocessor. The Norwegian Norcontrol Corporation successfully developed an integrated navigation system named the Data Bridge System in 1969. The system was mainly composed of four subsystems: data position, data radar, data sailing and data pilot. It is actually an integrated system based on collision avoidance and navigation.

IBS with Comprehensive Information Display and Automatic Track Maintenance (1970s~1980s)

In the 1970s, the shipping industry took measures of staff cuts to increase efficiency and at the same time vigorously promoted economic routes. With the successful development of adaptive autopilot, modern ships have put forward higher requirements for ship integrated driving control. The comprehensive control function of ship bridge systems has been expanded continually, the degree of automation has been improved greatly, the basic idea and configuration of IBSs are constantly improved, and various brands of integrated bridge systems have appeared. The Norwegian Norcontrol Corporation developed the DB2 data bridge in 1975 and the DB4 and DB7 data bridge in the early 1980s. Meanwhile, the U.S. Sperry produced its own IBS, the Soviet Union produced the IBS Wind-1, and the Federal German ATLAS Company produced the IBS NACOS-20. The development of IBS at this stage is mainly to increase the navigation plan and track maintenance autopilot function so that the basic automatic navigation of ships can be achieved.

IBS with the Fusion of Radar Images and Electronic Chart Information (1980s~1990s)

This phase witnessed further improvement in the IBS functions. The Norwegian Norcontrol company launched DB-2000, the British Racal-Decca company launched Miran 3000, 4000, and 5000IBS, and the German company launched the NACOS-25. At this stage, new ECDIS equipment was installed on the IBS system. As a result, this kind of IBS can superimpose radar video or radar tracking targets on ECDIS, and then the ECDIS integrates all kinds of navigation comprehensive information for drivers and displays the overall traffic situation to realize route control. A prominent feature of IBS at this stage is the realization of data transmission between subsystems through computer LAN, with the functions of navigation, control, route execution, management, communication and integrated display.

IBS with the Comprehensive Processing of Modern Navigation Information and Supervision of Navigation Safety

In the late 1990s, with the wide application of modern control theory, satellite technology, network communication technology and information processing technology, various information-based and intelligent new navigation instruments started to appear, and the IBS system also shows the tendency of diversified navigation information. Drivers are struggling to cope with a variety of navigational information from different sources. Therefore, IBS should have the comprehensive processing ability of navigation information to provide reliable, effective and precise navigation information to drivers. Comprehensive processing and information fusion of all kinds of navigation information is a remarkable feature of the IBS at this stage. For example, through comprehensive information processing, many functions are realized, including positioning, navigation, collision avoidance, self-driving, navigation management, communication, extinguishing and protection, life savings, simulation training, and alarm. It has also achieved the automatic control of bridge and cabin equipment and exercised the informative and intelligent management of ships. With international attention to the safety of life at sea and the protection of the environment, it is particularly important to monitor the normal operation of ship equipment and to supervise the driver's proper status on the bridge. On 20 May 2002, the IMO issued a performance standard for the MSC128 (75) Bridge Navigational Watch Systems (BNWAS). Although the standard is not enforced, some classification societies require IBS to be equipped with BNWAS.

At present, the world's large manufacturers of nautical instruments have the production capacity of the IBS, for example, US Sperry Marine company affiliated with Northrop Grumman Corporation, US Raytheon Corporation, British Kelvin Hughes Corporation, British Transas Corporation, German STN-ATLAS Corporation, Norwegian Norcontrol Company, Norwegian Kongsberg AS Company, Italian Consilium Selesmar Company, Japanese Company, Japanese TOKIMEC INC Company, Japanese Company, etc.

While the structure and composition of various types of IBSs by different manufacturers are different, and the classification methods of each subsystem of the IBS are different, the basic principle is the same: to combine many equipment and systems through electrical and mechanical combination, and the combination not only realizes equipment collection but also realizes function integrated and information integrated processing. All kinds of navigation information are utilized on a multifunctional workbench to realize the integrated control of the bridge.

Currently, although IBS is not obliged to be installed like other marine equipment or systems such as compass, radar, GMDSS, AIS, and VDR by the SOLAS Convention, it is the future of ship automation. After the IBS is put into use, the real ship application shows that IBS significantly reduces the ship track deviation and voyage increase rate, saves navigation time and fuel consumption, and as a result greatly improves the economic benefit. At the same time, it minimizes the driver's work burden and can ensure the ship's navigation safety to the maximum extent.

Future development of IBS will shift from traditional data acquisition and processing to decision-making and control. Increasing attention has centered on research on network technology, information processing technology, navigation expert systems, optimal route design, navigation integrated control, ergonomics and human–computer interfaces. Because of the characteristics and advantages of the IBS, it became the most energetic ship automation technology, and is also the main development tendency of shipboard technology.

Appendix 10.1: Questions

1. How can we accurately understand the basic concept, composition and historical development of the ship integrated navigation system? What are the basic configuration, functions and key technologies of the integrated bridge system?
2. Please briefly describe the relationship between integrated navigation system and integrated bridge system.
3. Please briefly describe the basic concept, technical development, projection type and application characteristics of the electronic chart system.
4. Please briefly describe the types, functions and technical characteristics of marine radio navigation AIDS.
5. Please briefly describe the composition and function of various parts of the navigation management system.
6. The main ship marine hydrological measurement equipment and meteorological element measurement equipment are listed, and the functions and characteristics of each piece of equipment are explained.
7. Please briefly describe the functions, connections and differences of heading autopilot and track autopilot, and explain the function and principle of P-I-D autopilot
8. Navigation techniques are used to guide the carrier to the destination, but this destination is usually determined externally. In addition, think about the interrelationship between realization by science and technology and selection by sense of value.

Appendix 10.2: Navigation and Value

Navigation technology can guide the carrier precisely to the destination, and the destination, however, is usually determined by the outside world. For example, decisions such as which port to be arrived are often decided by the captain. Since the route may lead to success, and may also be full of danger, navigation technology truly reflects the attribute of technical means. Similarly, science and technology can also be regarded as a means. The main goal of human science and technology is to meet the needs of human beings, but the establishment of this goal goes way beyond the pure science and technology and is determined by the ideology of people, such as values. This is a fundamental issue for individuals, countries and humans, and an issue often overlooked by researchers.

Ren Zhengfei (任正非, the founder of the Huawei company) once said: Technology is for social good. This statement illustrates the scientific and technological value of the modern Chinese civilization. When reviewing the research laboratories and centers around the world, one may recognize many other technological values in humanity. Due to the different values of science and technology, there will be completely different directions of technological development and, consequently, they may lead human beings to different directions.

   In addition to scientific technology, science is the main method for people to explore and understand the truth at present, mainly represented by formal logic and experimental empirical evidence. However, thus far, science has mainly managed to explain what the world is and why it is so, and failed to give satisfactory answers to questions about value. Since the nineteenth century, many western philosophers have been reflecting on the human factors behind "value" and "meaning", where the evaluation criteria also vary in different civilizations, countries, and cultures. Therefore, to cultivate scientific ideas, we should also have a better understanding of the boundaries of science. In other words, reason alone is never enough, and it should be value-driven. That is why we should first take a stand and then know good and evil. That is why it is believed that science has no borders, but scientists have a motherland. That is why we should avoid this narrowness and pursue the collective destiny and shared future of mankind.

References

1. Zhang Huijuan, 2014. Navigation from Sinan to Beidou [M]. Shanghai:Shanghai Science Popularization Press (62)
2. Li Yuese, 2006. History of Ancient Chinese Scientific
3. Zhao Lin, Yang Xiaodong, Cheng Jianhua, 2015. Modern Ship Navigation System [M]. Beijing: National Defence Industry Press (70)
4. Zhao Lin, Cheng Jianhua, Zhao Yuxin, 2011. Ship navigation and positioning system [M]. Harbin: Harbin Engineering University Press (70)
5. Liao Yongyan, 2007. Principles of Earth Science [M]. Beijing: Naval Publishing House (19)
6. Perloff AИ, Halysov BH, 2016. Principle of GLONASS Satellite navigation [M]. Liu Yining, Jiao Wenhai, Zhang Xiaolei, Liu Ying, translate. Beijing: National Defence Industry Press (29)
7. Paul D. Groves, 2015.Principles of GNSS and Inertial and Multi sensor Integrated Navigation System [M]. Lian Junxiang, Tang Kanghua, Pan Xianfei, Zang Kaidong, translate. 2. Beijing: Arms industry Press
8. Yang Jun, Dan Qingxiao, 2013. Principles and Applications of Satellite Time Service [M]. Beijing: National Defence Industry Press (50)
9. Gao Zhansheng, 2007. Fundamentals of Navigation [M]. Dlian: Dalian Naval Academy
10. Zhou Yongyu, Xu Jiangning, 2006. Ship Navigation System [M]. Beijing: National Defence Industry Press (76)
11. Bian Shaofeng, Ji Bing, Li Houpu, 2016. Introduction to Fundamentals of Satellite Navigation [M]. Beijing: Surveying and Mapping Publishing House (1)
12. Yuan Jianping, class, 2009. Principles and Applications of Satellite Navigation [M]. Beijing: China Aerospace Publishing House (56)
13. Li Yue, 2008. Navigation and Positioning—The Big Dipper in Information Warfare [M]. Beijing: National Defence Industry Press (17)
14. Fu Li, Wang Lingling, 2017. Concise Navigation System Tutorial [M]. Beijing: Publishing House of Electronics Industry (9)
15. Wu Dewei, 2015. Navigation Principles [M]. Beijing: Publishing House of Electronics Industry (44)
16. Sun Dajun, Zhen Cuie, Zhang Jucheng, class, 2008. Development and Prospects of Underwater Acoustic Positioning and Navigation Technology [J]. Journal of the Chinese Academy of Sciences, 2011, 34(06): 331–338
17. Zhen Cuie. Research on the Application of Ultra Short Baseline Positioning Technology in Underwater Vehicle Docking [D]. Harbin Engineering University, 2008
18. Wu Miao, class, 2015. Principles of Radio Navigation and Signal Reception Technology [M]. Beijing: National Defence Industry Press (45)
19. Zhang Wei, 2017. Principles and Methods of Astronomical Navigation for Deep Space Exploration [M]. Beijing: Science Press (64)

© Science Press 2024

H. Bian et al., *Essentials of Navigation*, https://doi.org/10.1007/978-981-99-5636-4

20. Xing Fei, class, 2017. Principle and Implementation Method of APS CMOS Star Sensor System [M]. Beijing: National Defence Industry Press (47)
21. Wang Hongli, Lu Jinghui, Cui Xiangxiang, 2015. Starlight Guidance Technology and Application of Large Field of View Star Sensors [M]. Beijing: National Defence Industry Press (36)
22. Klein I., Gutnik Y., Lipman Y. Estimating DVL Velocity in Complete Beam Measurement Outage Scenarios [J]. IEEE Sensors Journal, Volume 22, Issue 21, Pages 20730-7, 2022
23. Li Wen-tao, Qiu Wei, Zhao Ei-liang, Wang Chang-hong. Testing techniques of acoustic correlation log [J]. Technical Acoustics, Volume 34, Issue 4, Pages 362-7, Aug. 2015
24. Su Zhong, class, 2010. Inertial Technology[M].Beijing:National Defence Industry Press (34)
25. Xu Jiangning, class, 2009. Principles and Applications of Gyroscopes [M]. Beijing: National Defence Industry Press (48)
26. Yang Xiaodong, class, 2014. The Principle and Application of Semi—analytic Inertial Navigation for Ships [M]. Beijing: National Defence Industry Press (52)
27. Chen Yongbing, Zhong Bin, 2007. Principles of Inertial Navigation [M]. Beijing: National Defence Industry Press (6)
28. Britting KR, 2017. Analysis of Inertial navigation system [M]. Wang Guochen, Li Qian, Gao Wei, translate. Beijing: National Defence Industry Press (77)
29. Sun Wei, 2014. Spin Modulation Strapdown Inertial navigation system [M]. Beijing: Surveying and Mapping Publishing House (35)
30. Gao Wei, Beng Yueyang, Li Qian, 2014. Initial Alignment Technology of Strapdown Inertial navigastion system [M]. Beijing: Science Press (11)
31. Zhang Guoliang, Zeng Jing, 2008. Principles and Technologies of Integrated Navigation [M]. Xi'an: Xi'an Jiaotong University Press (61)
32. Wang Xinglong, Li Yafeng, Ji Xinchun, 2015. Integrated navigation technology [M]. Beijing: Beijing University of Aeronautics and Astronautics Press (40)
33. Bian Hongwei, Li An, Qing Fangjun, class, 2010. The Application of Modern Information Fusion Technology in Integrated Navigation [M]. Beijing: National Defence Industry Press (2)
34. Xu B., Guo Y. A Novel DVL Calibration Method Based on Robust Invariant Extended Kalman Filter [J]. IEEE Transactions on Vehicular Technology, Volume 71, Issue 9, Pages 9422-34, 2022
35. Ling Zhigang, class, 2016. Unmanned Aerial Vehicle Scene Matching Assisted Navigation Technology [M]. Shaanxi: Northwestern Polytechnical University Press (20)
36. Liu Ying, Cao Juliang, Wu Meiping, 2016. Research on Unmanned Aerial Vehicle Geomagnetic Assisted Positioning and Integrated Navigation Technology [M]. Beijing: National Defence Industry Press (26)
37. Zhang Xiaoming, 2016. Theory and Practice of Geomagnetic Navigation [M]. Beijing: National Defence Industry Press (65)